QUANTUM HORIZONS

CARL J. PRATT

Quantum Quill
Press

Table of Contents

Introduction

Greetings, cosmic voyager! Prepare to buckle up for an astronomical adventure as we delve into the subatomic reaches of reality. You're about to embark on a journey through *Quantum Horizons: Navigating the Secrets of the Universe with Quantum Physics*.

Quantum Horizons diverges from the traditional structure of science books. Instead of delving into complex math or following the conventional academic path to explain the intricacies of quantum physics and its relationship with our universe, this book offers a refreshing perspective. While my previous work, *Quantum Physics for Beginners*, followed a more classical approach; *Quantum Horizons* ventures differently.

As we dive headlong into the quantum realm, you'll find this is no ordinary physics book. We won't get bogged down with many abstruse equations. No, our venture eschews those typical detours, focusing instead on linking the phenomena of quantum physics directly to the awe-inspiring mysteries that embroider our cosmos.

Why this deviation from a more conventional academic path? This approach stems from feedback from readers, many of whom have expressed an eagerness to delve into the captivating aspects of quantum physics without being encumbered by its foundational intricacies. In essence, *Quantum Horizons* hopes to bridge the gap between curious minds and the wonders of the quantum realm, providing an inviting entry point for further exploration.

You see, every riddle we untangle is a cosmic question that's long tickled the human imagination, from the vanishing act of black holes to the tantalizing possibility of outrunning light itself. You might wonder: "Can we really extract energy from the elusive antimatter?"

Or perhaps you're puzzling over the true color of our blazing Sun. Maybe the enigma of white holes keeps you up at night.

With quantum physics as our compass, we'll seek answers to these and many more profound questions. Each chapter of this voyage has been crafted to discuss a different cosmic curiosity, all without resorting to mind-bending mathematics. This book, fellow traveler, is for you—the inquisitive, the curious, and those hungry to know more about our universe's mystifying fabric.

It's for those who yearn for more than dry, complex equations; those who seek to feel the thrill of understanding, the exhilaration of unveiling cosmic secrets. It's for anyone who's ever looked up at the stars, entranced by their silent ballet, and wondered, "What's really going on out there?"

In our quest, we'll traverse the bizarre landscape of quantum physics, where particles can be in two places at once, and cats can be both dead and alive—until observed. It's a realm that obliterates intuition, prompting us to reframe our notions about the universe's very nature. But don't worry, our journey won't require a Ph.D. All you need to pack is your curiosity and a thirst for the cosmos.

Together, we'll brave the bewildering and alluring depths of the quantum realm, where the very secrets of the universe are concealed. Remember, every great journey begins with a single step—or, in our case, a quantum leap. So, are you ready to uncover the universe's quantum conundrums? To grapple with the paradoxes that punctuate the cosmos? Then jump aboard this spaceship of discovery!

For beyond these pages lie the hidden mysteries of our universe, waiting, just as you are, for their secrets to be unfurled. Welcome, dear voyager, to *Quantum Horizons*.

CHAPTER 1

Parallel Universes: Fact or Fiction?

To understand if parallel universes are possible, we must start with the basics of quantum physics. Quantum mechanics offers us a profound insight into the behavior of particles by providing a means to calculate their future actions using the Schrödinger equation. This equation allows us to determine the evolution of a particle based on its quantum state, represented by its *wave function*. If you're not familiar with the concept of a wave function, or if you need a refresher, don't worry. Here's a quick explanation of what it is and how it works.

Think of the wave function as a recipe that tells us about the behavior of tiny particles, like electrons, in the quantum world. It's not something you can see or touch; it's a set of instructions written in the language of mathematics. The wave function is like a map of probabilities. It doesn't tell us exactly where a particle is, but it gives us the odds of finding it in different places. Imagine it like a weather forecast, which tells you the likelihood of rain in different areas, but doesn't guarantee where and when it will rain.

In our everyday world, it's impossible for something to exist in two places at the same time. However, in the quantum realm, particles can exist in multiple states at once, a phenomenon known as *superposition.* The wave function in quantum mechanics describes all these possible states.

Here's a strange thing—the act of measuring or observing a quantum particle changes its state. Before you measure it, the particle is in a blend of possibilities. But once you measure it, the wave function collapses, and the particle picks one definite position.

In quantum physics, you can't know everything precisely. For instance, if you know exactly where a particle is, you can't know exactly how fast it's moving. This fuzziness is built into the wave function.

Tiny particles in the quantum realm exhibit a unique dual behavior, acting both like waves and particles. This dual nature is encapsulated by the wave function. At times, these particles behave like ripples spreading across a pond, while at other times, they resemble individual raindrops.

The mathematics of wave functions in quantum mechanics employs a special type of number known as *complex numbers.* This is akin to using 3D glasses in a movie; just as the glasses add another dimension to our viewing experience, complex numbers add an additional dimension to how we describe the behavior of quantum particles.

In short, the wave function is a quantum recipe that tells us about the likelihood of where particles might be and how they might behave, but only in probabilities. It's a fundamental part of understanding the weird and wonderful world of quantum physics, where things don't always behave as we would expect in our everyday experience.

However, it is important to note that in the realm of quantum mechanics, we never directly observe the wave function itself, as depicted in the following picture.

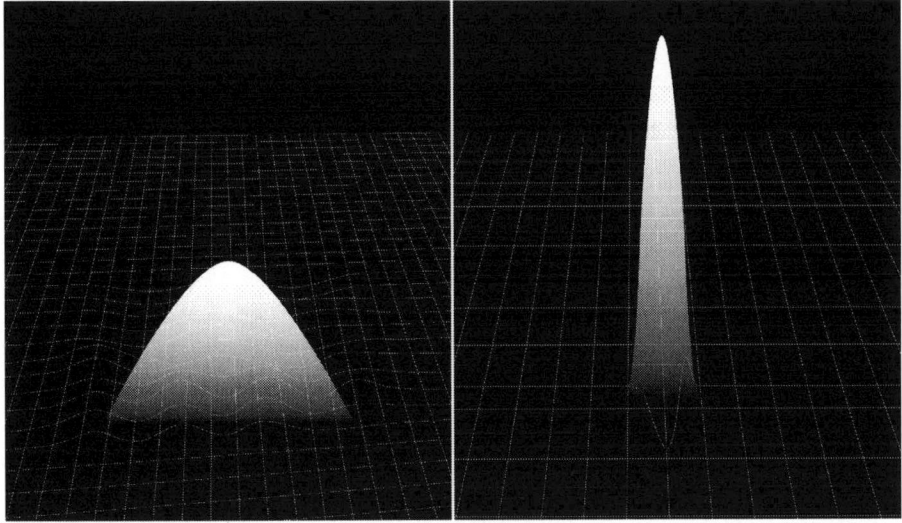

Fig. 1: On the left, a wave function represents the probability of a particle being in a specific position. On the right, the wave function collapses upon measurement, pinpointing the known position of the particle.

Still, upon measuring it, we discover the particle is located at one specific point in space. This raises a perplexing question: How do we reconcile the spread-out wave function nature with the particle's localized detection during measurement?

Indeed, the early pioneers of quantum theory understandably viewed measurements as more concrete and real compared to the elusive nature of the wave function. This perspective was rooted in the fact that measurements represented direct observations, aligning with our familiar experience of a world made of tangible matter particles. The outcomes of these measurements were easier to grasp and relate to, as they mirrored our everyday interactions with the physical world.

Our story has three leading players: Schrödinger, De Broglie, and Max Born. Think of De Broglie as the dreamer, the one who started suspecting that maybe, just maybe, matter isn't just stuff you can touch, but it also has a rhythm, a wave-like dance. Schrödinger was the craftsperson. He took De Broglie's dream and transformed it into a mathematical masterpiece—the wave equation.

But do you know what's funny about masterpieces? Sometimes, even their creators don't know what they truly mean. That's where Max Born enters the scene. He looked at Schrödinger's equation and saw not just waves but a map. A probability map. You see, the wave equation gave numbers for each point in space, and Born suggested that if you take these numbers and square them, you could predict the probability of finding a particle at that point.

Now, this was a wild idea. And it came as a footnote, quite literally, in Born's paper. But this footnote was a game-changer. It tossed the old view of a predictable, deterministic universe out the window and made room for a universe that deals with probabilities: no more certainties, only chances.

And you know what? This made some scientists, even Einstein, a bit queasy. But despite the philosophical turbulence it created, Born's rule has stuck around. Why? Well, because it's spectacularly good at predicting experimental results. That's the thing about the universe: it has a knack for surprising us, often in the footnotes.

In quantum mechanics, we have not one but two sets of rules. Here's how I usually put it: In moments of non-observation, the universe aligns with the smooth, continuous flow of Schrödinger's wave equation. However, this changes the instant an observation is made. At that point, when a measurement disrupts the system, the predictable melody alters dramatically. The result of the measurement becomes a matter of chance, influenced by a key feature of the wave equation.

One might assume that Schrödinger would appreciate such an unconventional interpretation of his equation, but surprisingly, he did not. In fact, his reaction was more akin to that of a displeased cat, an analogy he famously employed to express his dissatisfaction.

Picture this: a cat's inside a box (don't worry, it's just a thought experiment) alongside a radioactive atom and a very specific poison-release mechanism. This mechanism is hooked up to a detector. If the atom decays and radiation is detected, cyanide gas gets released and, sadly, our feline friend meets its maker. If the atom doesn't decay, no radiation is detected, no poison is released, and the cat lives to see another day.

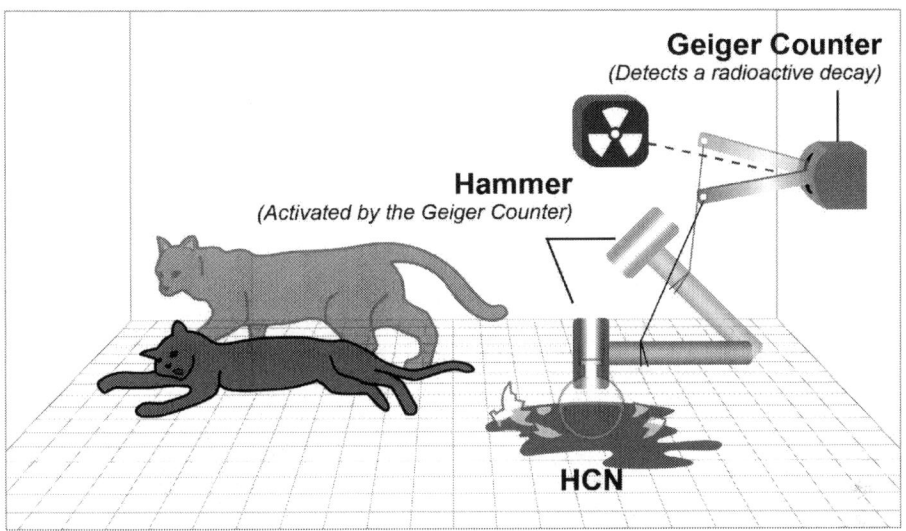

Fig. 2: Graphical representation of the renowned Schrödinger's experiment, featuring a cat inside a sealed box.

Now, Schrödinger didn't pick a cat because he hated them. He could have picked up any object. The point was to bring quantum weirdness to our everyday, macroscopic world.

At this point, the situation becomes quite intriguing. Quantum mechanics suggests that the atom inside the box isn't constrained to a single state of either decayed or not decayed. Rather, it can simultaneously exist in both states. This is akin to being in two places at once, a concept known as *quantum superposition*. However, the narrative takes an even more fascinating twist from here. This phenomenon of 'existing simultaneously in two states' extends beyond the atom. Indeed, it propagates to other elements in the system, including the detector and, as one might anticipate, the cat.

So, if you were to describe the situation inside the box, you'd end up with a head-scratching sentence: "the atom has both decayed and not decayed, the poison has both been released and not released, and the cat is both alive and dead." In the realm of quantum physics, this is a common scenario.

The crucial point is this: the cat's ambiguous state, alive or otherwise, is determined only upon observation. At that moment, the wavefunction, this quantum narrative, undergoes a dramatic shift, resulting in the cat being conclusively either alive or, conversely, not alive.

We often tell the tale of Schrödinger's cat as a quirky anecdote to underscore the downright bizarreness of quantum mechanics. But here's the real deal: Schrödinger wasn't aiming to entertain with this thought experiment. No, he wanted to yell, "Hey folks, something's off with quantum mechanics as we know it!"

Perhaps a different perspective could provide a clearer understanding, one that doesn't necessitate the concept of zombie cats. To achieve this, a deeper exploration is required. We must delve beneath the surface, examining our assumptions for any potential flaws in reasoning. Prepare for a journey into the depths of quantum theory as we prepare to reevaluate our current understanding.

Do you recall the mention of superposition from earlier? This concept has become crucial now. Superposition is the intriguing notion that a quantum object, like an electron, can exist in multiple states simultaneously. You might find yourself questioning the plausibility of this idea, thinking it seems far-fetched. However, the concept is not as implausible as it initially appears. Let me elaborate.

Are you familiar with the double-slit experiment? Let me explain how it works: we fire individual electrons at a screen with two slits. What we see is not just electrons taking one path or the other. Nope. We see an interference pattern, as if the electrons are playing both parts, going through both slits at the same time. Imagine an electron doing a dance through both slits simultaneously. That's superposition for you.

This might sound less peculiar if you think about waves. Picture waves spreading out, peaking and troughing. An electron behaving like a wave can do something similar. It can peek through one slit and trough through the other, leading to that fascinating interference pattern we see.

When nobody's looking, electrons act like waves. They're described by a wave function. So, our double-slit experiment is really a backstage pass to watch superposition in action. It shows us that an electron wave can indeed take two paths at once. It's a bit mind-bending, but in the quantum realm, superposition is standing on solid ground.

To progress with our reasoning, I need to briefly introduce the concept of *quantum entanglement*. We won't delve deeply into it; for now, it's enough to explore what it does at a high level and understand some implications of this curious phenomenon. This concept will reappear in subsequent chapters, and we will gradually build upon it as needed.

Picture this: two electrons zooming toward each other, moving at equal speeds but in opposite directions. We know they're going to bounce off each other, but the exact way they do that? That's up to the universe. Their paths are given by spread-out wave functions that deal with probabilities, not certainties.

Here's where it gets wild: the instant we determine the velocity of one electron, we immediately comprehend the velocity of the other. It must be equivalent but in the opposite direction to maintain momentum conservation. Simple, right? Well, there's a twist. Before we made the measurement, both electrons' velocities were in a superposition of states. Yet, measuring one instantly caused the other's wave function to collapse. Even if the electrons were light-years apart, this would still hold.

The truly fascinating aspect here is that post-interaction, the electrons no longer have their individual wave functions. They are instead orchestrated by a single-wave function. That's the essence of entanglement. And that's why measuring one electron instantly impacts the other—the shared wave function collapses. That's *quantum entanglement* in a nutshell.

There are clearly more formal and in-depth ways to explain this phenomenon, but in the context of this chapter, we just need to scratch the surface. There wouldn't be much benefit in overloading the reader with information that we don't immediately need.

You know, if we're getting philosophical about it, we could say there's just one big wave function—the wave function of the entire universe. It's kind of like an all-inclusive cosmic symphony.

Now, if we're addressing quantum particles that are independently operating, not entangled with anything, we can discuss their separate wave functions. However, the instant they engage with something else, the concept of entanglement enters the scene.

So, what we've got here is this: superposition describes systems in terms of waves, and entanglement tells us that interacting particles share a wave function. These are foundational pieces of quantum physics, characterizing systems with wave functions that dance to the tune of the Schrödinger equation.

But we've got one more player to discuss: measurement. Remember, the measurement postulate was added to bridge the gap between the math of the quantum world and what we actually see. But doesn't it strike you as odd? Why would there be one rule for when quantum systems are left to their own devices, and a different rule for when we're watching?

Breaking it down to the fundamentals, measurement can be understood as the interaction between quantum systems, such as electrons and photons. We already have a method to manage this interaction: allowing their wave functions to follow the Schrödinger equation. This process is straightforward. Therefore, the question arises: why should there be a different rule during observation? This presents a significant puzzle in quantum mechanics.

Let's consider a different approach. Imagine discarding the conventional rules of measurement and revisiting Schrödinger's cat scenario. The radioactive atom, existing in both a decayed and non-decayed state, becomes entangled with the detector and, subsequently, with the cat. It's important to recognize that we, as humans, are composed of electrons and atoms, and thus, we are also subject to quantum mechanics. This means we are as integral to the quantum process as the cat in the experiment.

When we open the box, there's no dramatic moment of measurement, no collapsing of the wave function. Instead, we just partake in the quantum dance and become entangled with the state of everything inside the box. This results in an odd twist in the narrative: we perceive the cat as both alive and dead.

I know what you're thinking: "How can that be? I've never seen a zombie cat!" Well, the answer lies in quantum theory's wildest suggestion. The version of you who sees the living cat and the version of you who sees the dearly departed cat are actually living in separate realities. Yes, you heard it right. These versions of you inhabit their own independent worlds, and these worlds, just like parallel lines, will never meet.

This is where it gets even crazier. Where do these separate realities come from? We need to bring in the silent players of this quantum game—the particles in the environment, the air molecules, the whole shebang that we've been ignoring.

Here's what happens: if a quantum entity in a superposition becomes entangled with this environmental crowd, it experiences a process known as *environmental decoherence*. This event forks the universe's wave function, splitting it into two slightly diverging realities.

Let's revisit Schrödinger's cat with this in mind. Our radioactive atom, initially in a state of being 100% not decayed, evolves into a superposition of both decayed and undecayed states. The detector then gets entangled with this split personality of the atom. But hold on; the detector isn't alone in there. It's constantly bombarded by air molecules and other particles inside the box, which react differently if the detector has picked up radiation versus if it hasn't. As a result, almost instantly, the detection device within the box becomes intertwined with the state of the environment and undergoes environmental decoherence. When such an event occurs, the wave function bifurcates into two.

At this very moment, there's a quantum photocopying effect: you are duplicated into two identical copies, each tangled with a different outcome of the experiment. These two versions of you stay identical until you open the box. Now, by the time you open the box, the cat is definitively either alive or dead, and the outcome is discovered only by opening the box. However, what we often fail to comprehend is that the alternate result also took place, but it happened to someone who is no longer you. Both observers, one who witnessed the cat alive and the other who saw it dead, originated from you, but they no longer represent you, and they've ceased to be identical. Quantum mechanics, indeed! It's a field that transcends mere atomic interactions; it delves into the division of realities as well!

So this intriguing way of looking at quantum mechanics is known as the *many-worlds interpretation*, a brainchild of the physicist Hugh Everett. If we buy into this version, then this wave function forking, this splitting of reality, is a continuous phenomenon. Heck, it could even be happening infinitely often, spawning countless, subtly distinct universes. Now, that might sound like we're veering into sci-fi territory, but remember, all these alternative worlds are already a part of quantum mechanics. They're embedded right there in the math. The many-worlds interpretation just takes them seriously.

To do away with these worlds, you'd need something akin to the collapse of the wavefunction. But here's the thing: your day-to-day reality would look and feel the same whether you're living in a multiverse or if wavefunctions collapse. But the many-worlds scenario offers a much cleaner, more elegant picture of the universe—it's all just wave functions dancing to the rhythm of the Schrödinger equation.

This leads us to a rather radical thought: the founders of quantum theory might have got it backward. The wave function isn't just a simple mathematical instrument; it embodies the entirety of reality. Our measurements and observations are merely small fragments, slices of the grand scheme that we become entangled with when we interact with a quantum entity in a state of superposition. This also brings us back to the realm of a deterministic universe. Every possible outcome happens 100% of the time in some reality or another. The randomness, the uncertainty we see, well, that's because we're only experiencing our tiny corner of the multiverse. So, the universe isn't just playing dice; it's playing dice in every possible way, in every possible universe, all the time.

Whenever I mention the many-worlds theory, I usually get hit with two big questions. Firstly, "Just how many worlds are we talking about, and how often do they split?" And secondly, "Does this mean anything and everything imaginable is happening somewhere?"

Addressing the first query, I must confess we're not quite sure about the exact number. It's a tad embarrassing, but the truth is, our knowledge in this realm is a bit hazy. However, let's be clear: we're talking about a LOT of worlds, a staggering number that's hard to even imagine. The theory goes that every time a quantum particle is in a superposition and interacts with its surroundings, it creates a split in the universe. For example, the atoms in your body decay 5,000 times every second, each decay potentially causing a split. That's a mind-boggling number of splits happening in just your body every second! Now, you may ask, is the universe splitting infinitely? We can't say for certain yet. We're not sure if the total number of possible splits is infinite or finite. We're talking numbers that are colossal beyond comprehension, and yet they could be finite. The more nuanced aspects are linked to areas of physics that we are still investigating, such as cosmology, quantum gravity, and the overarching theory of everything.

Now, on to the second question. The many-worlds theory doesn't imply that every conceivable thing happens. It simply means that everything that can happen according to the rules of quantum mechanics, as given by the Schrodinger equation, happens. The Schrodinger equation, our guidebook to the quantum world, gives us possibilities, but not everything is allowed. For example, it's impossible to observe an electron transforming into a proton. Such an occurrence would violate fundamental laws, such as the conservation of mass and charge. Within the framework of quantum mechanics, the probability of this happening is essentially zero.

So, circling back to our original question: "Are parallel universes really possible?" Well, according to the many-worlds theory, the answer is a resounding yes! This audacious idea suggests that every time a quantum event occurs, the universe "branches off," creating a brand new reality. These branches form an unimaginably vast multiverse with countless universes where every possible quantum event has occurred.

While it might be hard to wrap our heads around, this theory provides us with an astonishingly elegant and self-consistent interpretation of the peculiarities of quantum mechanics. But as with all theories in science, it's subject to refinement and even replacement as we dig deeper into the mysteries of the universe. So, are parallel universes really possible? In the grand scheme of things, the many-worlds theory tells us to keep an open mind because the universe is far more enchanting and complex than we could ever have imagined.

CHAPTER 2

The Dance of Light and Gravity

In the previous chapter, we talked about the possibility of multiple universes. From now on, however, let's focus solely on the universe we are familiar with. This journey continues with an analysis of the cosmic ballet of light and gravity.

It may surprise you, but gravity, the unseen force that keeps your feet planted firmly on the ground, also has a say in the paths that photons take.

What is a *photon*? A photon is a particle of light that carries electromagnetic energy and has no mass. Imagine light as a flowing stream, and within this stream are countless tiny droplets. In the world of light, these droplets are called photons. They're the reason behind the brilliance of a sunny day, the luminance of a lamp, and the diverse colors on your TV screen.

Now, to understand a photon a bit more, let's break down its main characteristics:

- *Quantum* often signifies the tiniest slice or piece of something. In light's context, a photon is that minuscule slice, the very reason why topics like quantum mechanics or quantum physics come into the picture when discussing light particles.

- Photons have a peculiar trait. At times, they behave like minute, speck-like particles. Other times, they resemble undulating waves akin to those in a serene lake. It's a duality that defies common understanding.

- Think of photons as ethereal entities. They lack mass, meaning they don't have weight. But surprisingly, they possess momentum and can even exert a push (a concept harnessed in space tech like solar sails).

- In a vacuum, photons maintain a steadfast speed: light speed. That's an astounding 299,792,458 meters every second.

- Different energy levels in photons translate to the myriad colors we see. For instance, the cool blue shade has photons pulsing with more energy than the warm red hue. This energy resonates with frequency—higher the energy, higher the frequency.

- When photons collide with objects, a few outcomes are possible. They might reflect, like sunlight off a shiny object; get absorbed, offering warmth to our skin; or simply pass through, much like light through crystal-clear water.

- Beyond illuminating our surroundings, photons play pivotal roles in advanced tools. From precision lasers to harnessing solar energy to aiding medical imaging like X-rays, photons are everywhere.

In essence, when you think of a photon, envision a minuscule, vibrant droplet of light. It might be tiny, but it's foundational to our understanding of the universe and essential to numerous technological marvels.

Now, let's return to the main focus of this chapter: the interplay between light and gravity.

You can think of gravity as a cosmic puppeteer, subtly manipulating the movement of light through the vast expanses of space.

Scientists have more than one explanation for this phenomenon. At a glance, these explanations seem to tell different stories, but they all reach the same conclusion. So, is there a "right" answer to how gravity and light interact? Or is truth as elusive as a photon?

Let's dive a bit deeper into the tale. Albert Einstein stated in his *General Theory of Relativity* (1915) that gravity has a hold on light, bending its path. Interestingly, this captivating idea was accepted by the scientific community long before it had solid proof.

Let's go back to 1783, when John Michell, a British clergyman, was pondering the universe's mysteries. He theorized that light particles, once captured by the gravitational field of a very large star, might slow down, briefly stop, akin to taking a short break, and then fall back into the star's gravitational hold. This idea marked the earliest discussion of what we now understand as *black holes*.

Michell shared his theory with his friend Henry Cavendish, who expanded on the concept. Cavendish hypothesized that as a light particle passes near a massive object, its trajectory would be slightly altered. This marked the beginning of our understanding of the intricate interplay between light and gravity in the cosmos!

Michell and Cavendish, in their theories, made some leaps of faith that were, quite frankly, way off target. They hypothesized that light, despite being massless, could be influenced by gravity, causing it to decelerate and alter its path, similar to how objects with mass respond to gravitational forces. They fully embraced Newton's perspective on gravity, theorizing that light, akin to a cooperative team player, would conform to Newtonian principles and respond to gravity in accordance with these established laws.

This leads to a particularly intriguing question: despite their initial uncertainties, Michell and Cavendish's predictions turned out to be remarkably precise. We now know that the relationship between gravity and light is far more complex than what Newton had envisioned. The true challenge lies in grasping the reasons behind this phenomenon. Why does light interact with gravity in such a profound way?

Picture this: you're in an elevator. Suddenly, it starts to accelerate upward. You feel a bit heavier, right? That's the pull of gravity. Now imagine you're in that same elevator, but this time it's free-falling. You'd feel weightless, right? At the right speed, you'll feel like floating in space.

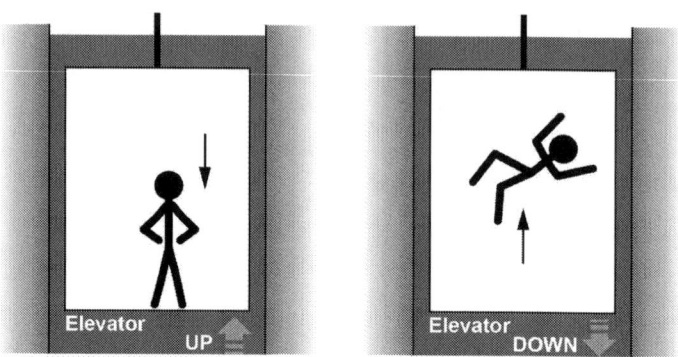

Fig. 3: On the left, the illustration shows the effect of gravitational pull on an elevator as it accelerates. On the right, it depicts the sensation of weightlessness experienced in a free-falling elevator.

This is the *equivalence principle of general relativity* (Einstein, 1915), stating that there's no difference between the sensation of backward pull when you're in an accelerating frame of reference and the downward pull of gravity of the same intensity. Alternatively, it can be said that the sensation of weightlessness experienced during free fall in a gravitational field is akin to the weightlessness one would feel if gravity were to, hypothetically, cease to exist for a moment.

When light ventures out of a gravitational field, it gets stretched out. This is what we call *gravitational redshift*. And here's the pivotal point: This prediction can also be made by acknowledging that time tends to progress more slowly within gravitational fields.

Any process creating a photon, a light particle, can be seen as a clock. It could be an electric charge bobbing up and down on a radio antenna or an atom shimmying in a light bulb because it's getting hot. These movements are like the tick-tocks of a clock. The speed of these movements (frequency) determines the wavelength and frequency of the photon that gets released.

However, when you're far away from this clock, it seems to slow down. This means the frequency of light escaping from a gravitational field is lower, and its wavelength is longer. If the gravitating body, the one creating the gravitational field, is dense enough, the light trying to escape can be stripped of all its energy and stretched out so much that its wavelength is virtually infinite.

Now, imagine you're standing on a sidewalk in bustling New York City. Out of nowhere, a cop car speeds toward you, its sirens blaring; the sound is piercing, high-pitched. But here's where it gets interesting. That siren—it's not constant. As that car zooms past you and starts getting further away, the sound of the siren drops in pitch. Lower and lower it gets, like a singer hitting those deep, soulful notes.

Why does that happen? Well, it's all about waves. When the cop car is speeding toward you, it's squishing those sound waves together, making them reach your ears more frequently. That's why you hear a higher pitch. However, as the car moves past and begins to distance itself, the situation changes significantly. The sound waves from the siren get stretched out. It's like the universe is pulling on either end of them, making them take longer to reach your ears. That is why you hear a lower pitch. The phenomenon described is known as the *Doppler effect*.

Fig. 4: A visual representation of the Doppler Effect, illustrating how sound waves are stretched or compressed depending on whether a police car is speeding toward or away from a person.

Now, replace that siren with light from a distant galaxy. When that galaxy moves away from us, its light gets stretched out, much like the sound waves from our cop car siren. Similarly, as the siren's pitch decreases, the light from the galaxy moves toward the red end of the spectrum. In the realm of light, red indicates a longer wavelength. This phenomenon is known as *redshift,* and it serves as the universe's method of signaling that an object is moving away.

Conversely, if the light source is moving toward the observer, the light waves are compressed, resulting in a decrease in their wavelength. This shift toward the shorter wavelength end of the spectrum, characterized by blue light, is known as *blueshift.*

Galaxy receding   **Galaxy approaching**

Light waves *stretched* - Red Shift   **Light waves *squashed* - Blue Shift**

Fig. 5: This illustration depicts two observers experiencing different phenomena due to the motion of galaxies. The observer on the left is experiencing redshift, where the galaxy moving away causes the light waves to stretch, shifting them toward the red end of the spectrum. Conversely, the observer on the right is experiencing blueshift, as the approaching galaxy compresses the light waves, shifting them toward the blue end of the spectrum.

Imagine you're near a black hole, a cosmic monster whose gravitational field is so intense that it puts time on hold at its event horizon. This is the infamous point of no return. Photons that attempt to break free from this gravitational grasp find their frequencies dropped down to absolute zero.

Black Hole Regions

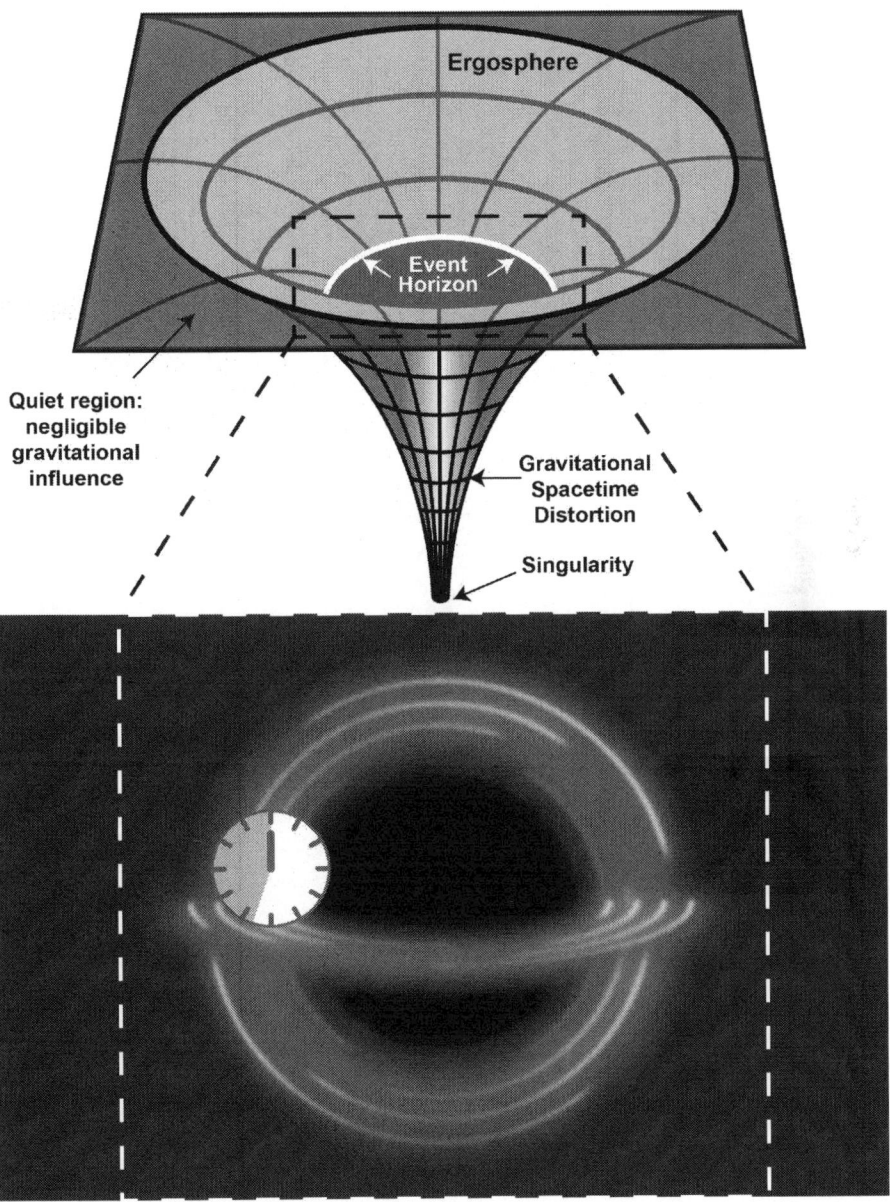

Fig. 6: At the top, the illustration depicts the various regions of a black hole. At the bottom, it shows the event horizon of a black hole and the effect of time dilation, where time appears to stop.

We have seen how gravity can stretch light's frequency, but what about how gravity bends the path of light?

For that, we turn to Huygens' wave theory of light, a game-changer in the world of optics. It proposes that a wave can be seen as an endless series of tiny, point-like oscillations, each spurring new waves. All these waves combined give us a pretty neat picture of the wave's future path.

Here's an easy way to visualize it: the ripples spreading out from the pebble can be seen as numerous new, smaller circular waves. As these smaller waves expand, they cancel each other out, except in the direction where the first ripple is moving. The result is a seamless extension of that initial ripple.

We can think about light in the same way, using the concept of plane wave of light to simplify how we model light behavior.

A plane wave of light is like a perfectly flat ripple spreading across a still pond. Instead of being curved or spreading out in a circle, it moves in one straight direction. In reality, light doesn't usually behave this simply, but imagining it this way helps scientists simplify and understand some situations.

This view helps explain why light can take detours around solid objects (a phenomenon called *diffraction*) and why it bends when passing from one medium to another (known as refraction).

Let's picture this: our plane wave of light is hitting the boundary of a new medium where light travels slower. If the wave approaches at an angle, the subsequent set of wavelets formed at the boundary will be more closely clustered together. They don't get as far in the time it takes for a new wavelet to form.

So, if we link these wavelets to reconstruct the overall wavefront, it becomes clear that the path of light has deviated. And there you have it—refraction! Now, it's a bit of a stretch, but you can think of light getting refracted or redirected by gravitational fields in a similar way.

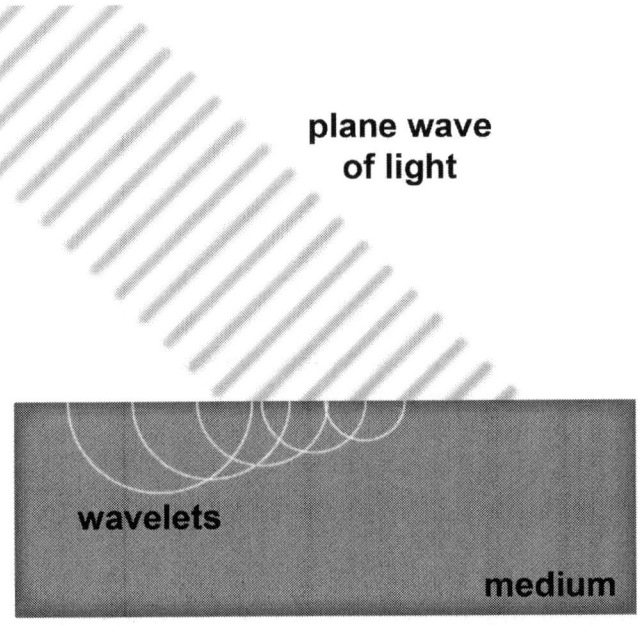

plane wave of light

wavelets

medium

Fig. 7: A plane wave of light hits a medium at a specific angle, where, in this medium, light travels more slowly.

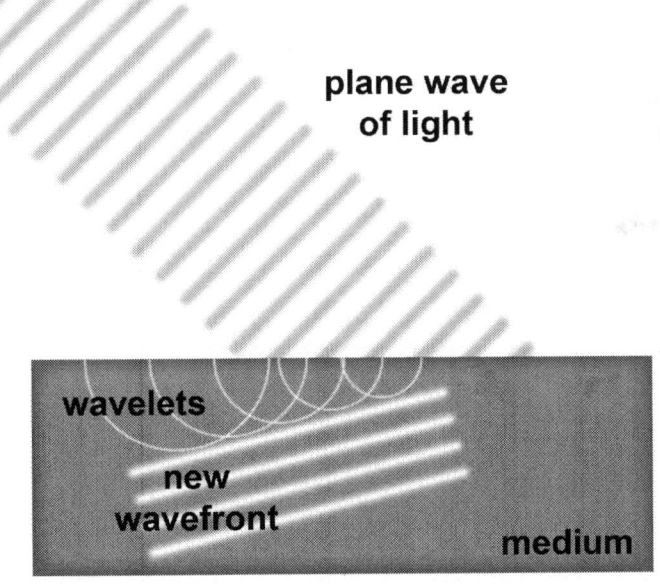

plane wave of light

wavelets

new wavefront

medium

Fig. 8: Formation of a new wave front following a deviated path.

For this, we need to embrace two pretty wild ideas. Firstly, that light behaves like a very old-school plane wave from the 17th century, which allows us to use Huygens' principle. Secondly, the speed of light shifts gears in gravitational fields, even though it seems to contradict everything we've talked about so far. Here's the intriguing part: this is precisely what Einstein proposed. Let's explore why this idea isn't as far-fetched as it might initially seem.

In the realm of relativity, as the term implies, most things are relative. However, there is one steadfast constant: the speed of light. It consistently travels at an unwavering speed, registering at 300,000 kilometers per second in a vacuum. Everyone, no matter where they are, observes this same local speed of light zipping through their own little patch of space. However, from a distance, the speed of light can appear to change.

Imagine observing Earth from afar. As a photon speeds past, you notice that it takes slightly longer to traverse the distance than it would if Earth were not present. This is because of two things: firstly, your clock is ticking faster compared to the clocks slowed down by Earth's gravitational pull. Secondly, space is stretched within the gravitational field, meaning the photon has more ground to cover through a time-slowed region. Both of these factors combine to make the speed of light appear slower.

Conversely, for an observer within the gravitational field, the photon appears to maintain its extraordinary velocity, moving at the speed of light as it passes by.

Even if this shift in speed isn't real, it's enough to put Huygens' principle into play. Think of each location perpendicular to a gravitational field as a vertical column of light, forming new wavelets at the wavefront. But, when observed from a distance, the speed of light seems to reduce nearer the gravitational source. This is attributed to the slowing down of time and the expansion of space in that region. As a result, from your perspective, the wavelets appear to cluster closer together. Now, if you trace the trajectory of the wavefront by joining these ripples, it appears to curve.

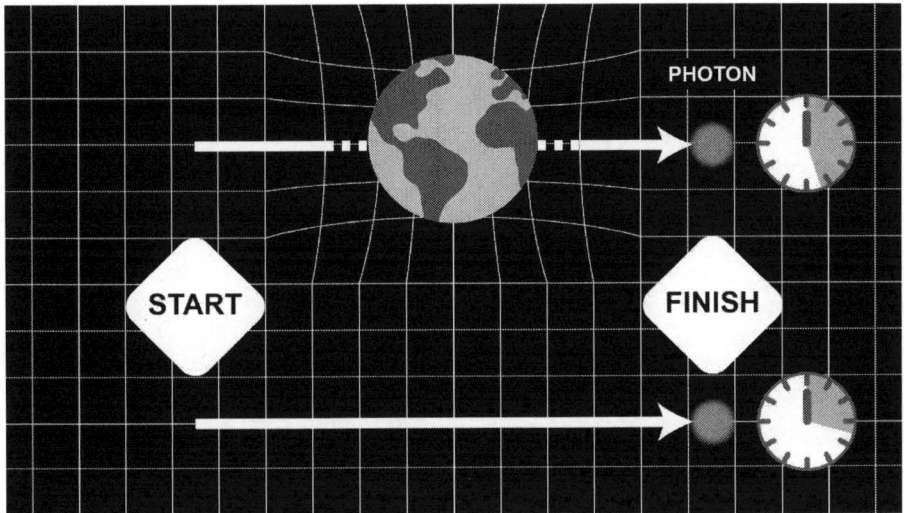

Fig. 9: How Earth's gravitational field influences the travel time of a photon from start to finish, compared to a photon not affected by Earth's gravitational field, as observed by an external viewer.

Einstein used this approach to calculate the deflection when light slips past a massive object. Interestingly, the deflection he calculated was exactly double that calculated by Cavendish, who used Newton's gravity theory.

In 1919, Sir Arthur Eddington embarked on a mission to put Einstein's prediction to the test. His plan was to capture images of the stars adjacent to the sun during a total solar eclipse. But he didn't stop there. Six months before this eclipse, he took additional photos when the Sun was far away from these same stars.

When he compared the two sets of star-studded snapshots, Eddington found something remarkable. As if following Einstein's script, the starlight did, indeed, bend in the sun's gravity.

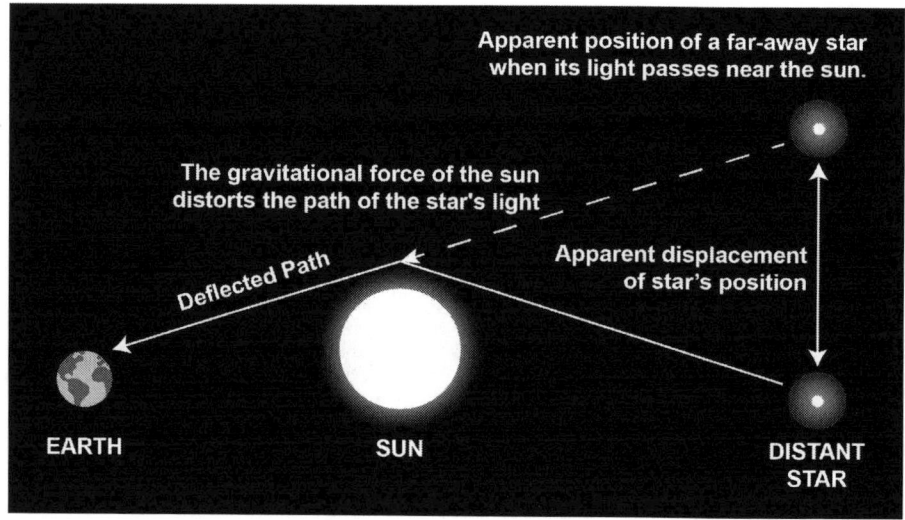

Fig. 10: The figures demonstrate how the Sun deflects the light coming from various stars before it reaches our eyes. This deflection causes the stars to appear in positions different from their actual locations.

Now arises a fundamental question: "Why was Einstein's calculation effective?" Light is not merely a straightforward plane wave; it embodies a unique combination of wave and particle properties in the quantum realm. The reality is that none of these representations perfectly capture the ultimate truth. It all boils down to perspective and how you define your reference frames.

In our universe, relativity is, fittingly, relative. The physical theories we formulate often lead us to discover that alternative, seemingly contradictory explanations can be equally valid. Light plays its dual role as a wave and a particle. Time slows, or space flows in gravitational fields. However, one steadfast rule remains: our universe exhibits a profound consistency within itself. Each of our explanatory tales is truthful in its own way, but they are merely slices of a more fundamental reality that underpins this generally relative spacetime we dwell in.

CHAPTER 3

Is It Possible to Outrun the Speed of Light?

Have you ever considered the idea that the universe might have a speed limit? A definitive boundary states, "This is the maximum speed attainable." If not, let me share this fascinating fact: the universe does have a speed limit, and it's defined by the speed of light.

If you're keen on physics, you might be thinking, "Wait a minute! Isn't this due to our mass increasing as we gain speed?"

Certainly! In the realm of physics, particularly when discussing the speed of light, there's a common belief that an object's mass increases as it accelerates closer to this ultimate speed. This concept, often associated with Einstein's theory of relativity, suggests that as an object moves faster, it gains more *relativistic mass*. However, it's crucial to clarify this point for a better understanding. The increase in "mass" is not about the object gaining physical matter. Rather, it's about how its energy increases as its speed approaches the speed of light. According to Einstein's equation $E = mc^2$, an object's energy (E) increases with its speed, making it require more and more energy to accelerate further as it gets closer to the speed of light. This exponential increase in energy requirement effectively prevents any object with mass from reaching, let alone exceeding, the speed of light. So, it's not so much about the object getting heavier in a traditional sense but about how the energy dynamics at high speeds make further acceleration increasingly difficult.

Why is it that we can't surpass the speed of light in this cosmic race? The answer is deeply rooted in a fundamental characteristic of our universe, and surprisingly, it involves some rather simple geometry. Allow me to explain this in more detail.

Einstein, with his extraordinary intellect, presented a revolutionary idea: space and time are not separate entities, like two children unwilling to cooperate. On the contrary, they are closely intertwined, acting as two aspects of a larger concept known as *spacetime*.

By integrating this insight with Einstein's theory of special relativity, we gain a deeper understanding. Einstein's theory introduces the concept of a universal speed limit, a velocity that remains constant regardless of the observer's location or their own velocity. Now, you can show that this speed is the speed at which particles without mass zip around. And since, as far as our current understanding goes, light particles are massless, we usually say this constant speed is the speed of light.

What if, one day, we discover that light particles do have a tiny bit of mass? Would that shatter Einstein's theory? Not at all. We would still have this constant speed; it just wouldn't be associated with the speed of light anymore.

With these concepts in mind, let's take a step back and use an analogy borrowed from Don Lincoln, a brilliant American physicist.

Picture this: a car cruising on an endless flat surface. But there's a twist: this car can only zip along at one speed, let's say 60 mph. And for our metric friends out there, that's 100 km/h. We'll overlay a simple compass, pointing north and east. Sure, we know the car's speed, but we're in the dark about how much of that speed is being channeled eastward and how much is shooting northward.

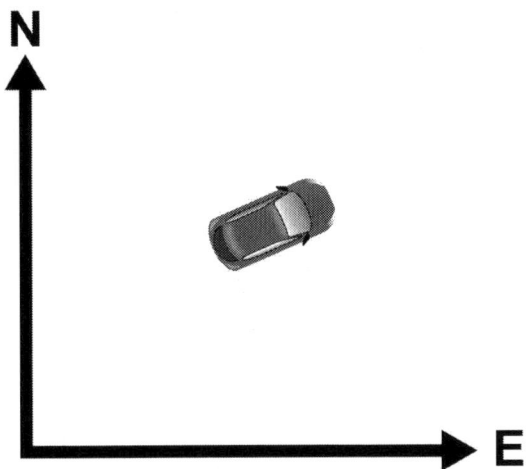

Fig. 11: A car cruising on an endless flat surface, with the X-axis indicating the East direction and the Y-axis indicating the North direction.

Suppose the driver decides only to head east. In that case, it's not making any progress northward.

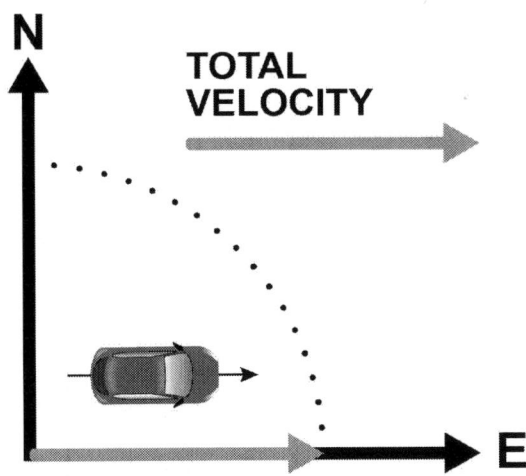

Fig. 12: A car heading eastward at its maximum velocity.

On the other hand, they could decide to venture north solely.

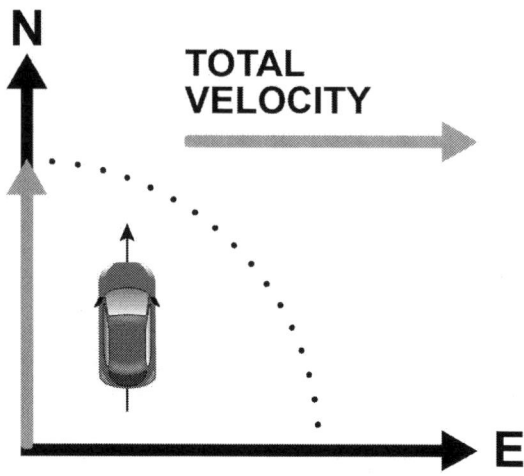

Fig. 13: A car traveling northward at its maximum velocity.

Or, the car could move in both the east and north directions, with neither claiming full motion.

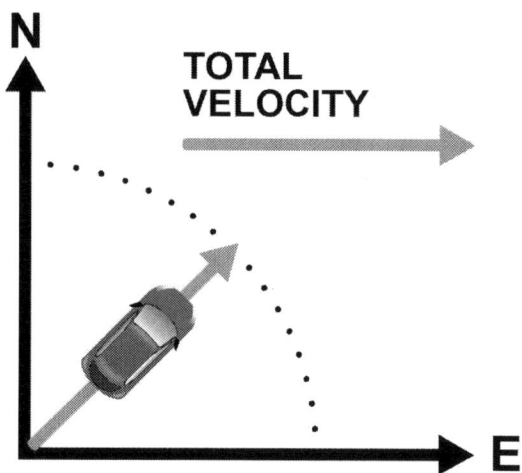

Fig. 14: A car traveling northeast, unable to reach its full velocity in either the north or east direction.

Now, let's bring in relativity in our analogy. In relativity, we don't have the east and north directions. Instead, we have spacetime. Let's imagine that the horizontal direction is space, and the vertical direction is time.

Consider this idea for a moment: what if there exists a singular, unvarying speed at which we can travel through spacetime? It sounds like a wild idea, but it's not far-fetched at all. Let's stir this idea into our car analogy from earlier. An object can decide to move vertically along the time axis; in that case, it's not going anywhere in space but cruising through time. If you're sitting at home or in your office and reading this, your position in space isn't shifting. Yet, you're living and experiencing time.

Now, let's flip the script. What happens when you start to dash through space? What if you pick up some speed? We find that the more you start to navigate through space, the less you journey through time. In fact, when you're moving solely through space, time stands still for you.

That's the crux of relativity. The quicker you go, the more your clocks hit the brakes. When you reach speeds close to that of light, your clocks freeze, almost like time itself is holding its breath.

So why can't we outpace light as we travel through space? It's because we're always gliding through spacetime at one speed—the speed of light. If we're not moving through space, we feel time zipping by at its fastest. But as we start to traverse through space, time starts to drag its feet more and more. Since we're navigating spacetime at this one speed, when we're moving only through space, there's no more speed to spare. We're blazing through space at the speed of light, and that's the maximum speed we are allowed to reach.

Now, let's give credit where it's due. Einstein wasn't the one who came up with this eye-opening realization. That honor goes to his mentor, Herman Minkowski, a man who could give Einstein a run for his money in the math department.

A mere two years after Einstein's ground-breaking 1905 paper, Minkowski grasped the geometric backbone of special relativity. He was the one who unearthed this profound reason why we can't break the cosmic speed limit set by light.

While Minkowski showed why light speed is the maximum speed through space, what he didn't explain was why we move only at one speed through spacetime. To this day, nobody really knows. It seems to be a fundamental property of spacetime. Perhaps the answer awaits discovery by a future mind as brilliant as Minkowski or Einstein.

CHAPTER 4

What Color Is the Sun, Really?

Have you ever wondered about the Sun's color? We often sketch it as yellow in our drawings, but the true shade of the Sun is quite intriguing—it changes based on numerous factors.

The sun actually emits light across a wide range of colors, from infrared to ultraviolet. When we view this spectacle from Earth, the light goes through our atmosphere, and this is where the magic happens. Our atmosphere is like a playful kid, tossing about the short-wavelength blue and violet light more than it does with long-wavelength red and yellow light. The result? The light that lands in our eyes is rich in warm hues like red, orange, and yellow, particularly during the poetic moments of sunrise and sunset. This is why the Sun often seems to have a yellow or orange tint to us.

But here's where it gets really fascinating! When we have the opportunity to observe the Sun from space or from high altitudes where the atmosphere is thinner, something magical happens. The sun appears brighter and takes on a closer-to-white hue. It's like a radiant beacon of pure light.

So, while we often label the Sun as yellow, it's more precise to say that it emits a whole spectrum of colors. Its perceived shade can change based on atmospheric conditions and the viewer's location.

Ever thought about how the Sun might look from the viewpoint of other planets in our solar system? It's a captivating question, and I've delved into some research to quench my curiosity!

Let's begin a cosmic journey, starting with our solar system's innermost planet—Mercury. The Sun, as seen from Mercury, is a vastly different sight compared to our earthly view. Because Mercury is nestled closer to the Sun, the star appears three times as big in its sky as we see from Earth. That's only the beginning. With Mercury being a mere one-third of Earth's distance from the Sun, the intensity of the sunlight there is a staggering spectacle—an observer on Mercury would see the Sun as seven times brighter than we do. Imagine that brilliant blaze!

Intriguingly, due to Mercury's absence of an atmosphere, the Sun's color would have an almost white appearance. Without an atmospheric curtain to play with the sunlight, the full spectrum of light becomes visible, casting a radiant white glow.

Here's a quick fun fact: One solar day on Mercury lasts a whopping 176 Earth days. Can you fathom that stretch of daylight? Light galore, but possibly a tad too much!

Our next stop is Venus, our system's second planet and famous for its thick, cloudy atmosphere. Alas, on Venus, glimpsing the Sun or stars isn't an easy feat. The Venusian sky is perpetually cloaked by a dense veil of clouds, efficiently blocking most light and clouding our view of the Sun. So, star gazing on Venus might not be the best idea.

Yet, even under such challenging viewing conditions on Venus, we can sketch a portrait of the Sun's appearance through the incredible images snapped by various space probes. Piecing together these visual breadcrumbs, we might picture the Sun donning a unique orange coat, thanks to the sunlight getting scattered and filtered by Venus's dense atmosphere.

An intriguing twist about Venus is its retrograde rotation—it likes to do its own dance and spins in the opposite direction to Earth.

So, if you were to plant your feet on Venus, you'd see the Sun rise in the west and set in the east, an astonishing contrast to our earthly experiences.

Next, let's jet off to Mars, the Red Planet, known for its unique atmosphere and rusty terrain. Mars boasts a thin atmosphere that, when combined with its dusty surface, creates an intriguing spectacle of light.

Though the Sun's luminosity on Mars is only 40 percent of what we're used to on Earth, the Martian surface isn't dim. Thanks to its thin atmosphere allowing sunlight to spread widely, the landscape is relatively bright. However, don't let that brightness deceive you—Mars isn't a warm place. With average temperatures hovering around a chilly -80 degrees Fahrenheit (-60 degrees Celsius), its feeble atmosphere struggles to retain heat, making it quite the frozen desert.

On Mars, the spectacle of sunrises and sunsets offers a delightful twist. Thanks to the *Purkinje effect*, the Sun is framed by a blue halo, while the rest of the Martian sky takes on a rosy, reddish glow. This presents a stark contrast to our experiences on Earth, where the sky near the Sun during sunrise or sunset is painted red. Mars's unique atmospheric composition and dust particles play artists here, creating a one-of-a-kind panorama of the Martian sky.

In addition to these mesmerizing lighting displays, standing on Mars, you'd notice that the Sun appears about 60 percent the size it does on Earth, a side effect of the differing distances from the Sun.

As we journey further to planets like Jupiter or Saturn, we don't have snapshots taken close to their surfaces. Nevertheless, we can assume the role of cosmic detectives, utilizing the data we have to piece together an image of the Sun's appearance.

Envision floating in Jupiter's upper atmosphere. The Sun would look quite different. Thanks to Jupiter's staggering distance from the Sun, the solar disk would shrink to a quarter of its earthly size. Additionally, the sunlight would be significantly dimmer, roughly 25 to 30 times fainter than what we bask in on Earth.

And the color? You might see a Sun with a blueish twinkle. That's thanks to Jupiter's unique atmospheric cocktail of gases and particles that interact with sunlight differently than our atmosphere does, potentially painting the Sun a subtle shade of blue.

But, plunge deeper into Jupiter's atmosphere, and the Sun becomes a faint memory. Dense, multicolored clouds cloak Jupiter's sky, swallowing the Sun into the whirlpool of vibrant hues—a truly alien environment.

Now, let's swing by Saturn. In this setting, the Sun takes a backseat to the planet's captivating rings, which become the primary spectacle in Saturn's sky.

From Saturn's surface, the Sun would be a fainter presence—about 100 times dimmer compared to Earth, thanks to the planet's greater distance from the Sun. Size-wise, the Sun would appear roughly ten times smaller than what we're used to. So, on Saturn, the Sun takes a humble bow, letting the rings steal the show.

But the real magic show in Saturn's sky is the play of light between the Sun and Saturn's majestic rings. As sunlight pirouettes through the complex ring system, it spawns amazing optical illusions. You might witness a *sundog*, where two dazzling spots flank the Sun, creating a celestial illusion of a triple star system.

Beyond sundogs, the rings can perform other tricks, like creating halos—a radiant circle that cradles the Sun, enhancing the celestial spectacle.

Journeying to Uranus, we find a world bathed in sunlight that has traveled an incredible two hours across the cosmos to reach it.

Because of this colossal distance, the Sun shrinks to a much smaller size when seen from Uranus—about 19 times smaller than our view from Earth.

Moreover, the faint trickle of sunlight that reaches Uranus plunges the planet into a softer, more subdued brightness. The Sun's glow is significantly dimmed here, providing only around 1/400th of the brightness we enjoy on Earth.

Pushing even further out to Neptune, the Sun undergoes another dramatic transformation. Because of the mind-boggling distance between Neptune and the Sun, from Neptune's surface, our star shrinks to a mere speck of light in the sky—about 30 times smaller than our view from Earth. And let's not forget, Neptune gets the short end of the solar stick among all the planets—it receives the least amount of sunlight. This sparse sunshine makes the Sun appear incredibly faint from Neptune, about 900 times dimmer than our earthly experience.

This short tour reminds us of the vastness and diversity of our solar system. It highlights the unique characteristics of each planet and how they shape our perception of the Sun. So, the next time you cast your gaze upon the Sun, take a moment to appreciate not only its ever-changing hues but also the scientific wonders that lie behind its captivating and colorful display.

CHAPTER 5

What's the Secret Behind Star Survival?

Imagine a cosmic game of tug-of-war, with our universe as the playground. On one end of the rope, we have the relentless force of gravity pulling everything toward a center point. On the other end, we have the energy from stellar fusion pushing everything outward. It's a celestial stand-off, and this delicate balance can persist for billions of years as if the universe held its breath.

So, what happens when a massive star, running low on fuel, allows gravity to finally claim victory? Well, that's when the cosmic playground witnesses a spectacle of unimaginable proportions. This overpowering of fusion energy by gravity can trigger an extraordinary collapse, erupting into a tremendous blast we call a *supernova*. What's left in the aftermath could be a neutron star or even a black hole. But why does such a rapid collapse generate such a powerful explosion? And why are neutron stars or black holes formed? Strap in; we're about to take a deep dive into the captivating tale of the birth of neutron stars and black holes.

First, let's get a grip on how a supernova comes into being. A star produces outward pressure to counteract gravity through the energy it generates from fusion. This fusion process takes us on a journey up the periodic table, starting with humble hydrogen merging to form helium. Then helium pairs up to create carbon, carbon produces neon, and so on, until finally, an iron-nickel core is formed. At this point, fusion reaches its limit because the energy required to form elements heavier than iron exceeds the energy it produces. This sets the stage for the spectacular event of a supernova.

Not every star is destined to forge iron in its core. Take our Sun for instance—it won't make it to this stage. It will exhaust its fuel before it can start producing iron. Yet, for a sufficiently massive star, its fusion process culminates in iron. At this juncture, the fusion reactions cease, halting energy production. This cessation results in a decrease in the outward thermal pressure. Gravity, ever-present and waiting, seizes this opportunity to initiate the star's collapse.

At this stage, the core contracts for a while, held in check by a phenomenon known as *electron degeneracy pressure*. Now, you might ask, "What's that?" Well, it's all about the *Pauli Exclusion Principle*. To explain it, let's introduce first the concept of *fermions*, in case you are not familiar with them.

Named after the physicist Enrico Fermi, fermions are a kind of elementary particle that makes up what we call matter. All the things you can touch, taste, or see? Made up of fermions. Particles like electrons, protons, and neutrons are all fermions. They're the building blocks of atoms and, hence, the building blocks of the universe.

Now, here comes the interesting part. Fermions, in the quantum realm, act as the ultimate rule enforcers. They adhere to what's known as the Pauli Exclusion Principle. This principle states that two identical fermions cannot exist in the same quantum state at the same time.

Put simply, this rule dictates that identical particles can't share the same location and energy state simultaneously.

A classic illustration of this principle can be seen with electrons in an atom. Being fermions, electrons must adhere to the Pauli Exclusion Principle, necessitating that they inhabit distinct energy orbits or levels when circling an atom's nucleus. The significance of this principle is foundational, underpinning the composition of atoms and, consequently, the fundamental stability and structure of matter.

It's like they each need their own personal bubble. This is why matter has volume and structure!

What does this imply? Typically, atoms have ample room to move and vibrate. However, when a star starts collapsing, gravity steps in like an overzealous bouncer, squishing the atoms together until they're in their most contracted state. What we end up with is the electrons in the outermost shell of the atoms forming a kind of barrier, keeping the atoms from getting any closer to each other.

Now, you might think, "Ah, this is because like charges repel, right?" But here's the plot twist: Freeman Dyson, a theoretical physicist, clarified in 1967 that this physical barricade is a result of the Pauli Exclusion Principle, not due to neighboring electrons or nuclei repelling each other. There are simply no other vacant quantum states for the electrons to move into. There's literally no way for them to compress further. This is the crux of electron degeneracy pressure. As the outward radiation pressure diminishes, this newly significant pressure comes into play, contending with gravity to prevent further collapse of the star.

Once a star hits this degenerate state, gravity meets its match. Quantum mechanics steps in, declaring "No Vacancy!" There's no more room at the inn for further compression. So, the core of the star manages to survive, not thanks to the outward pressure of fusion, but because of the quantum laws that keep it from fully collapsing.

This sturdy heart of a star, resisting gravity's immense force through electron degeneracy pressure, is known as a *white dwarf*. That's the fate that awaits our very own Sun in a few billion years. After it exhausts its nuclear fuel and casts off its outer layers during its red giant phase, it will leave behind a white dwarf: a dense, Earth-sized remnant with a mass 200,000 times that of Earth.

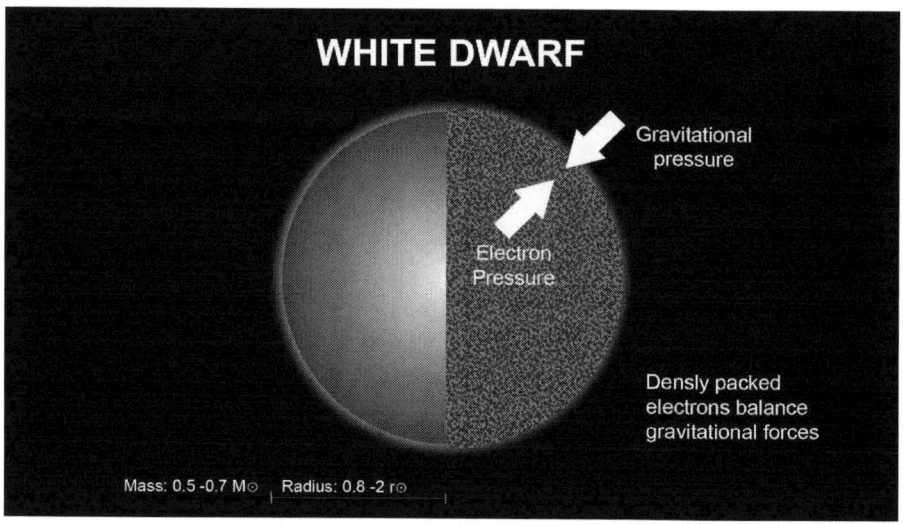

Fig. 15: How electron degeneracy pressure counteracts gravitational force, preventing a white dwarf from collapsing.

Now, to wrap your head around the enormity of that mass, consider this: Jupiter is merely 320 times more massive than Earth. The Sun is 300,000 times more massive than Earth but also 110 times wider. However, there's a catch: a white dwarf has a mass limit. Indian American physicist Subramanian Chandrasekhar figured out that this upper limit is about 1.4 times the mass of the Sun. This boundary is known as the *Chandrasekhar limit*.

If the white dwarf manages to stay under the Chandrasekhar limit, this degeneracy pressure—that's the *quantum-mechanical pressure*—continues to put up a good fight against gravity's relentless squeeze. What happens next is quite remarkable. The white dwarf, over an unimaginably long time, cools down, losing its heat bit by bit, until it ultimately freezes solid.

So, what happens when a collapsed star tips the scale at over 1.4 times the mass of our Sun? Even the mighty electron degeneracy pressure can't prevent further collapse. Extra mass can be piled on from the star's outer layers, pushing it over the Chandrasekhar limit. Once this boundary is crossed, there's no stopping the collapse. The star continues its catastrophic plunge.

When a star pushes past the Chandrasekhar limit, things get seriously hot. Core temperatures skyrocket to over 5 billion degrees Celsius. At these temperatures, the star churns out gamma rays—little packets of energy so potent they smash iron nuclei into alpha particles, also known as *helium nuclei*. This nuclear alchemy cranks the heat even more to an unimaginable hundred billion degrees Celsius.

In these intense conditions, electrons and protons are drawn together irresistibly and merge in a phenomenon known as *electron capture*, resulting in the formation of neutrons. This cosmic matchmaking also unleashes a tidal wave of tiny, elusive particles called *neutrinos*. So, once our star steps over the Chandrasekhar limit, its electrons get scooped up by atomic nuclei, transforming into a tightly packed ball of neutrons. Our star is no longer a star. It's become something eerier—a colossal atomic nucleus.

This neutron star is an even more compact package than a white dwarf. Forget being the size of Earth; it shrinks to a mere 10 to 20 kilometers wide. But the density? It's a mind-boggling 100 trillion times that of Earth! Picture this: a teaspoon of neutron star stuff would weigh as much as Mount Everest.

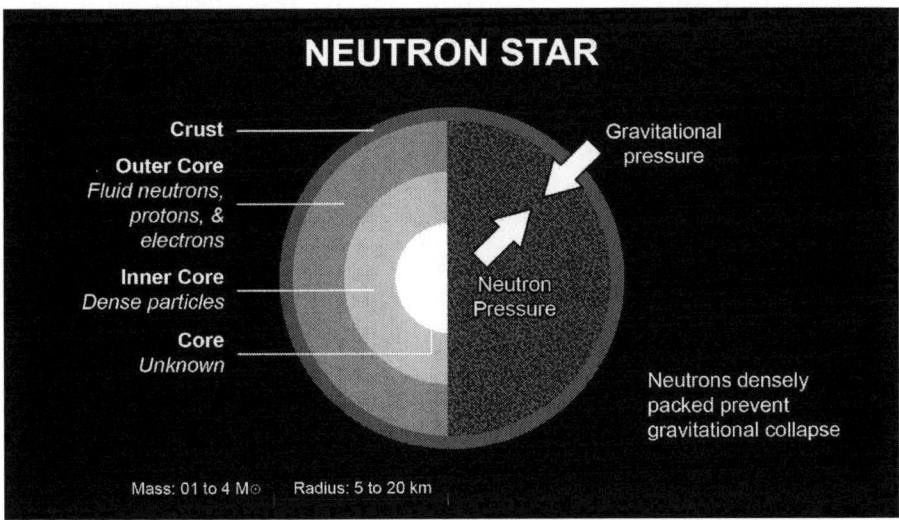

Fig. 16: How neutron degeneracy pressure counterbalances gravitational force, preventing a neutron star from collapsing.

Now, remember the Pauli Exclusion Principle that keeps white dwarfs from collapsing due to electron degeneracy? It turns out that the same principle applies to our neutron star. It's all because the principle applies to all fermions. Electrons are fermions, and neutrons, being composed of quarks, which are also fermions, fall under the same rule.

Let's assume our star continues to pile on mass, exceeding the Chandrasekhar limit; instead of the electron degeneracy pressure we had earlier, a *neutron degeneracy pressure* steps in. This means the star can't be squished any further—the neutrons are already as cozy as they can get. The core of the star, now more like a colossal atomic nucleus, forms an indomitable wall. As the star's outer layers collapse inward, they encounter this neutron barrier and are abruptly repelled. This rapid recoil generates a shockwave that surges through the star, propelling the outer material outward at escape velocity. And there you have it: the birth of a supernova!

This dramatic compression and subsequent outward bounce releases an insane amount of energy—up to 10^{46} joules within a mere 10 seconds. This spectacular energy show is what we spot with our telescopes, and sometimes even with our naked eyes, when a star in our own galaxy goes supernova. For a fleeting moment, a single supernova can outshine an entire galaxy.

The most recent guest appearance of a supernova in our Milky Way galaxy was back in 1604. This interstellar fireworks display popped up like a surprise bright star, even visible during the day. Chinese scholars noted this celestial event, and the astronomer Johannes Kepler studied it. The remnants of this supernova, discovered in 1941, still twinkle in our telescopes today.

To add variety, the universe presents us with several types of supernova explosions, each characterized by its own unique mechanisms and timelines. The scenario I've described falls into the category of core-collapse supernovae. These are somewhat akin to the 'vanilla' flavor in the spectrum of cosmic explosions, representing a common and fundamental type.

I walked you through the process step-by-step, which might've given you the impression that this takes a while. But the reality is that once electron degeneracy pressure buckles, the core implodes dramatically within seconds. With the support of the inner electron degeneracy core gone, the outer core collapses like a house of cards, hurtling inward at nearly 25 percent the speed of light until it smashes into a wall of neutrons—courtesy of neutron degeneracy pressure.

To put things into perspective, here's the timeline of events: A massive star may spend millions of years approaching the end of its life, but its core collapses in less than a quarter of a second. Following this, it takes just a few hours for the resulting shockwave to surge to the star's surface. The explosion itself is brief, lasting only about two minutes, but its peak brightness sticks around for several months, and it can take over a year to completely fade from view.

After this celestial spectacle concludes, what remains is a *neutron star*, an entity capable of persisting for billions of years.

But you might be wondering, if gravity can overpower electron degeneracy, what's stopping it from doing the same to neutron degeneracy? Good question, because it absolutely can.

Just like the Chandrasekhar limit has its star-crushing point, leading to a neutron star, there's a sibling limit, the *Tolman-Oppenheimer-Volkoff limit*—let's call it the *TOV limit*. If this TOV limit is breached, it results in the ultimate universe's vanishing trick: a *black hole*. Researchers in 1996 suggested the limit is somewhere between 1.5 to 3 times the mass of our Sun. So, if the leftover neutron star is heftier than that, gravity wins again, smashing through even the neutron degeneracy pressure to squish it further into something even smaller than a neutron star—potentially infinitesimally small. We're talking about forming a black hole—a place where gravity is king and every other force takes a back seat. It's a celestial object that pushes our equations to their limits.

When a black hole is born, it's the universe's ultimate no-exit zone. Nothing can break free, not even light. So, unless matter is falling into the black hole, heating up and compressing to unleash high-energy radiation like X-rays and gamma rays—visible to our telescopes—it's virtually invisible. That's why we call it a black hole.

So here's the takeaway: white dwarfs and neutron stars both owe their existence to the same principle—the Pauli Exclusion Principle —that says two fermions can't share the same state or position, creating a wall against further compression. The difference lies in the type of fermion. For white dwarfs, it's the electron. For neutron stars, it's the neutron.

CHAPTER 6

Black Hole Evaporation

Black holes truly capture the imagination as some of the most enigmatic entities in the cosmos. Their name is quite fitting, as they possess inherent darkness, absorbing all light that ventures too close. In essence, they appear as punctures within the fabric of space-time itself.

The surface of a black hole is known as its event horizon. It represents the boundary beyond which the escape velocity required to break free exceeds the speed of light. Once an object crosses this threshold, it becomes inexorably drawn into the gravitational grasp of the black hole. It is important to note that the space within the event horizon behaves unlike anything we typically encounter in empty space—it possesses a uniqueness all its own.

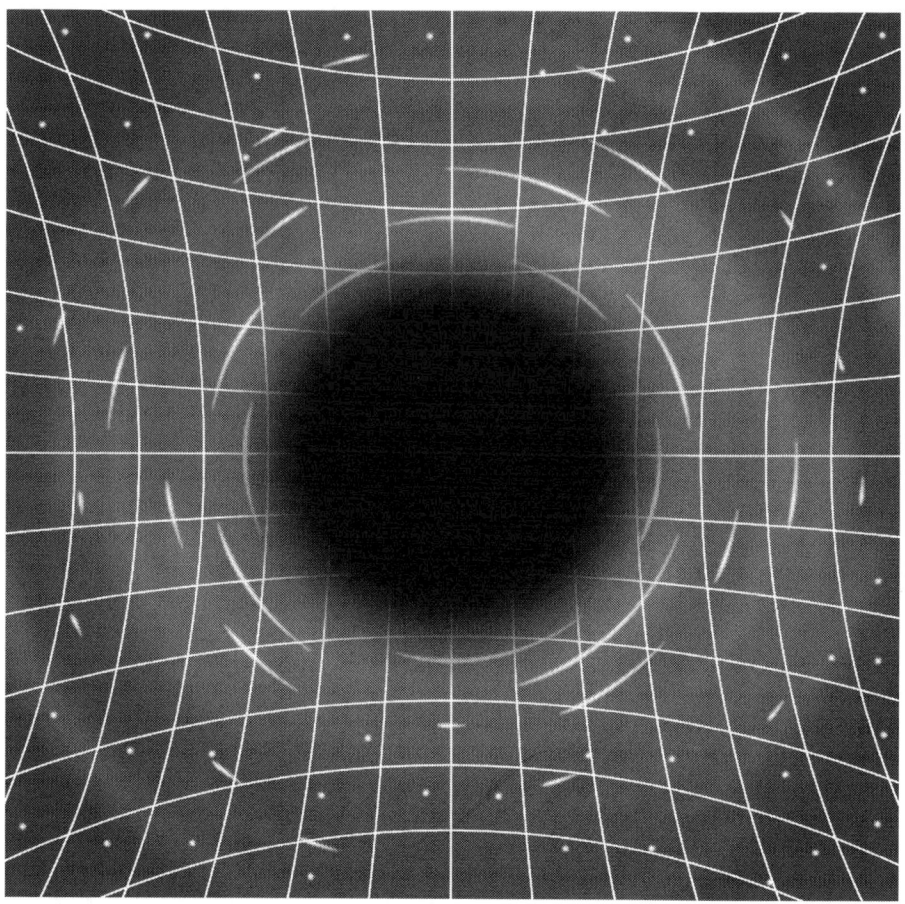

Fig. 17: Graphical representation of a black hole, including its event horizon.

In 1974, the brilliant physicist Stephen Hawking introduced a groundbreaking idea that challenged our conventional understanding of black holes. He employed quantum mechanics principles in his computations, integrating them with the traditional physics that had previously guided our understanding of these mysterious entities. What he discovered was truly remarkable—a revelation that black holes can emit radiation and, in fact, shine!

According to the principles of quantum physics, black holes have the ability to release photons, the fundamental particles of light. This notion might strike you as perplexing, given that black holes are renowned for their propensity to absorb all light. So, where do these emitted photons come from? How do they come into existence within the clutches of a black hole?

The gravitational pull of a black hole is extraordinarily powerful, fostering the common perception that it is inescapable, even for light. Without the lens of quantum mechanics, traditional physics posits that a black hole's mass cannot decrease, but rather stay constant or increase, given the presumption that nothing can evade its gravitational clutches.

To determine the surface area of a black hole, we can employ a straightforward formula utilized for any spherical object. This formula, $A = 4\pi r^2$, gives us the surface area of a sphere.

However, when it comes to black holes, the equation is modified. The key parameter in this context is the Schwarzschild radius, denoted as Rs. This radius is directly related to the mass of the black hole and can be calculated using the equation $Rs = (2GM)/C^2$.

In this equation, G represents the gravitational constant, M denotes the mass of the black hole, and C^2 corresponds to the speed of light squared.

The idea that a black hole's surface area could remain constant or increase fascinated Professor Hawking and his peers. This concept was particularly intriguing because it mirrored the second law of thermodynamics—an essential principle in studying heat and energy flow.

The second law of thermodynamics articulates that "In any natural process, the entropy of an isolated system either increases or, at a minimum, stays the same." Entropy gauges the level of disorder or chaos within a system, and according to the second law, this state of disarray typically escalates as time progresses.

Drawing inspiration from this fundamental principle, Stephen Hawking proposed what is now known as *the second law of black hole mechanics*. According to this conjecture, "In any natural process, the surface area of a black hole's event horizon is projected to either expand or, at minimum, stay unchanged. Remarkably, it will never decrease."

Similar to thermodynamics, we can find an analogous formulation in black hole mechanics, revealing striking parallels in their fundamental principles. Just as thermodynamic laws control the dynamics of heat, energy, and work within a system, the principles of black hole mechanics adapt these concepts to the distinct realm of these cosmic phenomena. These similarities not only deepen our understanding of black holes but also hint at profound connections between different areas of physics. Below are some critical parallels drawn between the two:

Zeroth law:

- *Thermodynamics:* Same temperature throughout equilibrium
- *Black hole mechanics:* Same surface gravity everywhere on the horizon

First law:

- *Thermodynamics:* Energy is conserved
- *Black hole mechanics:* Mass is conserved

Third law:

- *Thermodynamics:* Can't reach zero temperature
- *Black hole mechanics:* Can't reach zero surface gravity

The intriguing connection between black holes and the laws of thermodynamics leads us to ponder whether black holes might be akin to thermal objects. In the realm of thermodynamics, there exists a concept called a *black body*—a theoretical object that absorbs all incoming radiation without transmitting or reflecting any. Similarly, a black hole behaves in a comparable fashion, absorbing all radiation that encounters its gravitational grasp.

Herein lies the conundrum. If we consider a black hole as a black body, it would imply that it possesses a temperature associated with it. According to established principles, a temperature corresponds to thermal radiation being emitted. In other words, if a black hole has a temperature, it should emit some form of radiation, which seems to contradict classical physics' understanding that a black hole does not release anything.

This dilemma challenges us to reconcile these seemingly conflicting ideas. Is there a resolution? Who holds the correct perspective?

When Stephen Hawking embarked on his quest to understand the behavior of black holes, he initially found the idea of these cosmic entities emitting radiation to be highly implausible. Determined to explore this further, he ventured into the uncharted territory where quantum mechanics intertwines with general relativity—and the remarkable unfolded.

In 1974, Hawking published a seminal paper outlining a mechanism that revealed the astonishing truth: black holes possess the capability to shine. The explanation he presented is elegantly simple, yet it carries profound implications for our understanding of the universe.

He started from the assumption that the fabric of space itself teems with a continuous interplay of virtual particles—particles that momentarily come into existence before disappearing once again. This endless ballet, referred to as matter-antimatter annihilation, spans across the entirety of space and time, happening unceasingly. This is grounded in the *Heisenberg uncertainty principle*. One interpretation of this principle can be expressed as follows:

$$\Delta E \, \Delta t \gtrsim \frac{h}{4\pi}$$

$\Delta E =$ Uncertainty in energy

$\Delta t =$ Uncertainty in time

$h =$ Planck's Constant

This formula signifies that the uncertainty in energy and the uncertainty in a timeshare have an inverse relationship, as their multiplication results in a constant value.

In simpler terms, if we possess precise knowledge of a system's energy, we cannot have precise knowledge of the time over which the measurement was conducted, and vice versa. This suggests a basic boundary on the exactness with which these characteristics can be simultaneously determined.

However, the equation also unveils an intriguing possibility. It suggests that particles can exist with a certain energy range, represented by ΔE, as long as this energy fluctuation occurs for an exceedingly brief period of time, represented by Δt. Crucially, if the product of these uncertainties is less than Planck's constant divided by 4π, these particles can indeed exist, even in apparent violation of the uncertainty principle.

One of the mind-boggling aspects of quantum mechanics is that it allows for violations of conventional expectations, as long as they occur within a timeframe that evades direct measurement. It's almost as if the universe disregards these violations of the Heisenberg uncertainty principle, simply because no measuring device can ever capture them directly.

Consider a particle with a finite amount of energy. In the realm of quantum mechanics, this particle can exist for a brief period as long as the change in time is exceedingly small. This peculiar phenomenon arises from the concept of virtual particles, which can momentarily emerge from the fabric of empty space, borrowing energy from the surrounding environment. However, these particles swiftly annihilate each other, promptly returning the borrowed energy.

This dynamic interplay of particle creation and annihilation gives rise to what we refer to as *quantum foam*. It populates the seemingly empty expanse of space, teeming with virtual particles that briefly materialize before disappearing once more.

Yet, close to a black hole's event horizon, the severe warping of space-time alters this quantum foam in manners that deviate from the patterns seen in regular vacuum space. As particles and antiparticles spring from the vacuum, a unique circumstance can take place. If a pair of these particles come into existence near the event horizon, one particle might be pulled into the black hole before they can annihilate each other.

particle with positive energy

particle with negative energy

Fig. 18: A particle and its antiparticle spontaneously emerge near a black hole. One of them is captured by the black hole's gravitational field, preventing the pair from annihilating each other.

In this scenario, where one particle near the black hole's event horizon is captured, its partner is left to roam without a counterpart for the annihilation process. These particles, escaping the clutches of the black hole, carry energy with them and manifest as what we call *Hawking radiation*. This remarkable phenomenon provides an explanation for how black holes shine and sheds light on the origin of the emitted photons.

What's intriguing is that, from our perspective, the released particle has a positive energy. This implies that the black hole has acquired a particle with negative energy, resulting in a minuscule loss of energy for the black hole. In the context of Einstein's famous equation, $E=mc^2$, this loss of energy is equivalent to a loss of mass. Hence, we can interpret the process as the black hole gradually diminishing in mass.

To summarize, the concept of Hawking radiation suggests that virtual particles are generated in space by borrowing energy. However, to uphold the fundamental law of energy conservation, the energy that manifests as radiation is ultimately derived from the mass of the black hole itself.

While this explanation is widely embraced, it encounters a significant challenge. The radiation emitted by a black hole does not span across all wavelengths as one might expect with this mechanism. Contrarily, the radiation displays a unique trait: its wavelength is proportional to the size of the black hole. As a result, smaller black holes discharge radiation with shorter wavelengths, suggesting that they emanate more energy compared to larger ones.

Hawking approached the question of black holes' radiation from a different perspective, employing quantum field theory and wave analysis in his calculations. To provide a simplified explanation, let's avoid delving too deeply into the intricacies of quantum field theory.

Hawking's computations took into account waves originating from infinity and interacting with the black hole's gravitational field. When these waves met the black hole, some oscillations were diverted or modified due to the powerful gravitational impact. A portion of these waves was warped or entirely consumed by the event horizon, while others remained unaltered.

Hawking's groundbreaking insight revealed that the waves that managed to enter the event horizon underwent disruptive processes. As a result, the wave emerging from the other side of the black hole carried away energy proportional to the size of the black hole itself.

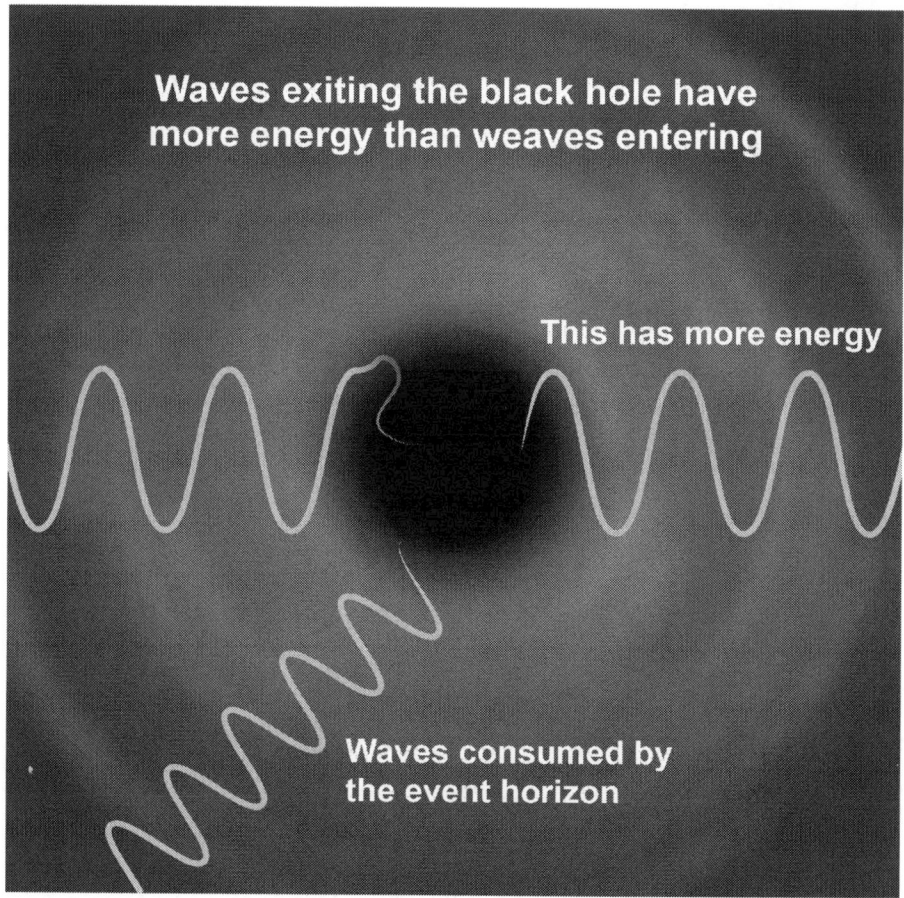

Fig. 19: Representation of the effect of a black hole on wave patterns.

Particles, represented by waves as expansive as a black hole's event horizon, get trapped in its gravitational grip. These waves, with wavelengths similar in size to the black hole, cannot break free and essentially vanish within the event horizon.

However, it is crucial to recognize that quantum fields associated with wavelengths on the scale of the black hole exhibit a distinct behavior. The waves absorbed by the black hole must possess negative energy, allowing us to observe the manifestations of positive energy within our universe. This intriguing phenomenon ensures a delicate balance between positive and negative energy within the quantum realm surrounding the black hole.

The waves released align with an energy spectrum similar to that of a black body at a specific temperature. This means that Hawking radiation is perceived as the temperature of the black hole itself.

Hawking radiation cannot be measured directly, but scientists found a formula for the temperature of a black hole.

$$T = \frac{x}{M}$$

$$x = \frac{hc^3}{8\pi Gk}$$

$T =$ Temperature of the black hole

$M =$ Mass of the black hole

$h =$ Planck's Constant

$k =$ Boltzmann's constant

$G =$ Gravitational constant

$c =$ The speed of light in a vacuum

It's important to note that as the mass of a black hole decreases, its temperature increases. In simpler terms, as a black hole "shrinks" or "evaporates," it does so at an accelerated pace. In fact, when the mass approaches zero, the evaporation rate becomes infinitely fast. This intriguing insight implies that during the final stages of the evaporation process, we would witness a dramatic event—an explosion of the black hole. As the remaining mass is rapidly consumed, this explosive phenomenon will manifest as a burst of intensely energetic particles known as gamma rays.

The question now is: how long does it take for a black hole to complete its evaporation process? Well, the lifetime of a black hole can be calculated using this equation:

$$t = yM^3 \qquad y = \frac{5120\pi G^2}{hc^4}$$

Upon performing the calculations, it becomes evident that the process of black hole evaporation is extraordinarily slow. In fact, a black hole of significant size would require billions of years to complete its evaporation.

COMPARISON OF THE EVAPORATION TIMES FOR BLACK HOLES OF VARIOUS SIZES

Mass in Grams	Equivalent object	Time to evaporate
10^8	20 elephants	1 second
10^{10}	Eiffel Tower	12 days
10^{15}	Mt. Everest	13.8 Billion Years
10^{33}	Sun	10^{66} years

Considering these findings in relation to the age of the universe, it leads us to speculate that in the immediate aftermath of the Big Bang, there might have been a population of exceptionally tiny black holes that underwent rapid evaporation. However, as billions of years have passed since the Big Bang, we are now reaching a point where we could potentially observe the evaporation of some of these smaller black holes. It's worth noting that the term 'small' here is subjective, as we are referring to black holes roughly comparable in size to Mount Everest.

If this hypothesis holds true, it would imply that our telescopes should be capturing numerous gamma-ray bursts, as it is theorized that a black hole would release an outburst of gamma rays before its ultimate disappearance. As expected, we do detect a substantial number of gamma-ray bursts. Since its monitoring starting in 2005, the NASA Goddard Space Flight Center has recorded an average of one gamma-ray burst per day.

However, the observed pattern of these gamma-ray bursts does not align with what we would anticipate from a black hole explosion. Rather than observing a seamless and progressive surge in brightness from a lesser to a greater value, ending in a final explosion, we witness bursts displaying variations in luminosity. These bursts alternate between periods of brightness and dimness before returning to brightness again.

The occurrence of these gamma-ray bursts is more likely attributed to other astrophysical phenomena, such as the collision of neutron stars or the explosive demise of supermassive stars, rather than the evaporation of black holes. As a result, the available data does not seem to support the notion that extremely small black holes exist, as we would expect to observe their distinctive gamma-ray bursts if they were present.

Although direct evidence of Hawking radiation remains elusive, it harmoniously aligns with the principles of quantum mechanics, and only a negligible number of physicists question its existence. The theoretical framework supporting Hawking radiation is widely accepted within the scientific community. Despite the lack of direct observation, the concept elegantly integrates into our understanding of the laws governing the quantum world.

CHAPTER 7

Do white holes really exist?

White holes are like the cosmic yin to the black hole's yang. But do they exist? Well, let's unravel this stellar enigma together.

Rewind to 1916, when Karl Schwarzschild, a mathematician of remarkable intellect, discovered one of the most straightforward solutions to Einstein's general relativity. He conceptualized the geometry of space and time surrounding a singular mass point. Imagine condensing a massive chunk of cosmic matter into an infinitesimal dot within the immense expanse of space.

In the model proposed by Schwarzschild, the space-time geometry features a horizon that acts as a one-way boundary—you can enter but never leave. Even traveling at the speed of light, it's impossible to reverse past this horizon. This makes it a point of no return, capturing light within and presenting itself as a dark sphere, what we commonly refer to as a black hole.

In the core of this space-time model, where all the matter is compressed, both the density and the curvature of space-time reach their extremities, becoming infinite. In the world of physics, we term this a *singularity*. Once you've stepped over the black hole's horizon, you're not just barred from heading back; you're also caught in a one-way tumble toward the singularity. It's like a cosmic pit you can't help but fall into.

In Schwarzschild's solution, crossing a black hole's horizon is akin to a journey from the past to the future. Once past it, you're always facing the singularity, being drawn toward it, much like time incessantly nudges us toward the next instant. It's a journey that's set in forward motion with no reverse gear.

But hold on—isn't this peculiar? Usually, our physics equations are time-neutral. They make sense whether you read them front to back or back to front. They don't favor the past or the future.

In the realm of physics, a significant number of essential equations exhibit a characteristic where time's direction doesn't alter their validity. This feature is referred to as *time symmetry*. To illustrate, consider a scenario where a planet's orbit around a star is recorded and then played in reverse. The gravitational principles governing the orbit remain consistent regardless of the time direction. This is also true for other phenomena, such as the interaction of particles; their behavior conforms to physical laws, whether time moves forward or backward.

This characteristic implies that the fundamental laws of physics don't inherently distinguish between what we perceive as the past and the future. These laws apply equally well whether time is progressing in its usual manner (from past to future) or in reverse (from future to past).

However, this notion of time symmetry contrasts with our everyday experience of time's unidirectional flow—from the past, through the present, and into the future, a concept often described as the *arrow of time*. This concept becomes evident, particularly in the context of the second law of thermodynamics. This law posits that an isolated system's overall entropy (or disorder) will tend to increase over time, introducing a distinct asymmetry in the flow of time.

In summary, while many fundamental equations in physics do not inherently prefer a direction of time, our observable universe, influenced by entropy and the arrow of time, appears to exhibit a preference for one direction over the other.

So, why this time-biased, singularity-bound journey? We've often said that the answer lies not in the equations but in the initial conditions and the statistical ballet of a multitude of particles.

So, how do white holes fit into this? Well, if a black hole is a one-way journey from the past to the future, could a white hole be the opposite—a one-way trip from the future to the past? Theoretical physics certainly permits such an idea, but the question is, do they exist in reality? That remains one of the universe's most captivating unsolved mysteries.

Time symmetry is an intriguing concept. It pops up in general relativity too, its equations don't dictate any fixed direction in time. So, you'd be justified in wondering whether our understanding of black holes is incomplete, whether there's a missing piece that could restore this symmetry.

The calculations needed to unravel this are pretty complex, but let's cut to the chase. On deep analysis of Schwarzschild's geometry, we find an exciting twist: there's indeed a solution that's perfectly time-symmetric to the black hole. Think of it as a mirror image in time, or a black hole played in reverse. This curious concept is aptly named a *white hole*.

Now, let's try to wrap our heads around this. A black hole, as we know, has a horizon that you can enter but never leave—it's a one-way cosmic door. What's the perfect symmetry to this? Naturally, a horizon that you can only exit but never enter. So, while a black hole's horizon sucks in everything, a white hole's horizon behaves like a radiant cosmic fountain. Instead of swallowing light, it spews out matter and radiation, appearing white to us—hence the name.

Moreover, with a black hole, once you cross the horizon, you're destined to meet an inescapable singularity in your future. However, for a white hole, the singularity lies in the past, always behind you. So, you can only originate from the singularity, but you can never head toward it.

Remember how, with a black hole, you can never actually see anything cross the horizon? From an outside perspective, time slows down as you near the black hole, making it seem like an object takes forever to cross the horizon.

In contrast, a white hole spits out light that takes an eternity to reach an outside observer. Thus, observing a white hole becomes a tricky business.

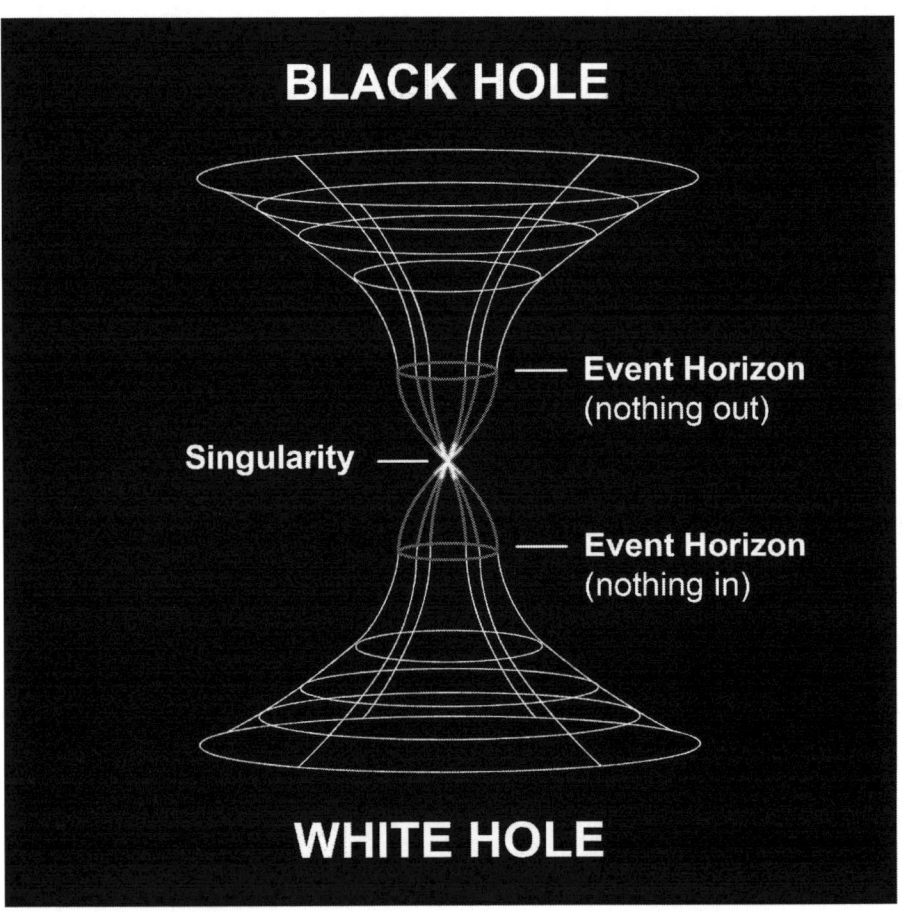

Fig. 20: A comparison between a black hole and a white hole.

We've mathematically defined what a white hole is, but the real question remains: can they actually exist? First off, let's remember that the Schwarzschild metric is an idealization. It outlines space-time in a vacuum around a point mass, that is, mass compacted into a dimensionless point. So, we're talking about a point with infinite density, a singularity. But does such a situation actually occur in our universe? That is still up for debate.

Yes, Schwarzschild's geometry is an idealized concept, but it still gives us a darn good approximation of the space-time around any spherical object like a star. We're confident black holes can form because if a spherical object collapses in on itself while remaining roughly spherical, Schwarzschild's geometry has got it covered. So, we have a real, cosmic process that can create a black hole, and indeed, we've found concrete evidence of their existence.

However, things get trickier with white holes. You see, unlike black holes, we haven't identified a physical mechanism that could form a white hole. Here's why.

Firstly, Schwarzschild's idealized solution, which brings black holes into the picture, is time-symmetric. It implies an eternal situation where a white hole has always existed to infinity in the past, and a black hole will exist forever into the future. But hold on a minute! Our universe has only existed for a finite time. There can't be eternal objects. Black holes can spontaneously form at a certain point, but an object that has existed from eternity past? That's a no-go.

Furthermore, when a black hole forms from, say, a collapsing star, the part of the solution describing a white hole simply doesn't exist. Reality isn't perfectly symmetric in time.

And here's another reason white holes raise eyebrows. Their formation would run counter to the arrow of time, breaking the second law of thermodynamics. While the laws of physics don't favor any temporal direction, spontaneous processes always head one way in the real world: toward increasing entropy.

A star collapsing into a black hole fits this bill. It's a process moving in the right direction, the direction of increasing entropy. No special conditions are needed. But to create a white hole in our universe? You'd have to reverse this direction and swim against the tide of increasing entropy. That's not just unlikely; it's pretty much considered impossible. This is why, even though they exist in the mathematical realm, white holes remain a controversial topic in the real world of astrophysics.

Let's put it this way: forming a white hole would be like piecing together a shattered glass into a whole one again. But in reality, the creation of a white hole would be an even more fantastically unlikely event.

Remember, we're still only discussing general relativity when we speak of these mathematical solutions. But in conditions near the horizon or the singularity, we should also consider quantum effects. The trouble is, we're still wrestling with marrying quantum mechanics and general relativity together. We've still got mysteries to solve.

Sure, there are theoretical studies in quantum gravity that speculate about the creation of white holes from black holes in our universe. But these are just educated guesses at the moment. We're not sure if they hold water.

Now, the only thing in our universe that even remotely echoes the concept of a white hole is the Big Bang. When we look at the simplest description of the Big Bang from general relativity, we see a singularity in the past that we can't reach, from which matter and radiation sprang.

The notion that our universe sprung from a white hole isn't totally off the table. Some theories even propose that every time a black hole collapses in our universe, it births a white hole, leading to the creation of a new universe. But we're nowhere near getting a letter from the other side, so we can't verify these claims. Maybe our universe came about through such a process, but right now, we have no concrete reasons to believe this or evidence to prove it.

So, here's the scoop: black holes might seem like sci-fi material, but we know they're out there. White holes are just as mind-bending, if not more so. But as of now, we don't have any proof of their existence, and there are substantial doubts if they can exist at all.

CHAPTER 8

The Matter Puzzle: Debunking the Myth of Four States

Have you ever wondered why your desk, coffee mug, or chair maintain their shapes? Well, it's all thanks to the trusty solidity of their chemical bonds. They're like a group of loyal friends, holding each other tight, making sure they don't lose their shape.

But here's where it gets interesting. Imagine turning up the heat—literally! As we do, the bonds holding our solid object together start loosening up, like friends at a party starting to dance. They move more freely and fluidly, and our solid becomes a liquid. It keeps its connections, like those dancing friends, but with a newfound freedom to move.

Keep cranking up the heat, and those bonds will relax even more. Our particles get so excited they break free, like birds flying out of a cage, and our liquid turns into a gas. These particles, once held tight, are now exploring their surroundings independently.

But we're just getting to the good part. Picture taking things to a whole new level—super high temperatures, like the heart of a star. This intense heat causes an extraordinary spectacle. Electrons, the particles that usually orbit atoms like planets around a sun, get kicked out. This breaks all the remaining bonds and leads to a state we call plasma. It's a dazzling display of free-roaming, charged particles—matter's own fireworks show!

Now, you might be thinking, "Solid, liquid, gas, plasma, got it. We've seen all there is to see," but hold on a moment! The universe, in its endless creativity, has much more up its sleeve. In fact, the scientific community has uncovered or theorized many more states of matter. It's like an ongoing cosmic treasure hunt, expanding our perception of the myriad ways matter can behave.

It would be quite a challenge to traverse all these states in our little exploration here, but we can have a look at the most interesting ones.

However, before we delve deeper, allow me to briefly introduce *quarks*, assuming you may not be acquainted with them.

Quarks are one of the smallest building blocks of matter, invisible to our most advanced microscopes and most sensitive detectors, but so essential to our existence that without them, the universe as we know it would crumble.

Physicists once believed that protons and neutrons, which constitute the nucleus of an atom, were the most fundamental, indivisible components of the universe. However, the universe had a surprise in store. As it turns out, both protons and neutrons are composed of even smaller particles known as quarks!

Now, quarks come in six *flavors*—but you can't taste them, unfortunately. The flavors are *up, down, charm, strange, top,* and *bottom.* Up and down quarks are the most common and are found in protons and neutrons. The other four are heavier and much rarer.

Quarks are never found alone in nature. They're always in groups, a phenomenon known as *confinement.* It's like they're always throwing a party, and nobody wants to leave. Physicists have never been able to separate quarks; believe me, they've tried.

The universe's fundamental structure seems to have a preference for balance, particularly in the case of quarks, the basic constituents of matter. Quarks possess a unique property termed *color charge*, which is a theoretical concept rather than a literal color. This concept dictates that the universe favors a state of color charge equilibrium.

Consequently, quarks commonly group in sets of three, as observed in protons and neutrons, with each quark in the trio having a distinct color charge—red, green, or blue—creating a balanced mix.

In another scenario, quarks can form pairs with their antimatter counterparts, known as *antiquarks*. In such pairs, the quark and its corresponding antiquark have opposite color charges, like a red quark pairing with an anti-red antiquark. This pairing leads to the neutralization of their color charges. These formations, either in trios or pairs, are integral for particle stability and the overall structure of matter. They are a part of the principles of color confinement in the field of quantum chromodynamics, which is the study of the strong force in particle physics.

So, quarks might seem elusive, bizarre, and downright strange. But they're an integral part of the cosmic symphony, the subatomic notes that help to compose the universe's grand opus.

Now, let's go back to the main topic, exploring new states of matter. Even within subatomic particles, such as quarks residing within protons and neutrons, we encounter unique forms of matter. This prompts us to ask intriguing questions: What states of matter do quarks exhibit? Do they depend on the states of matter of the atoms they belong to?

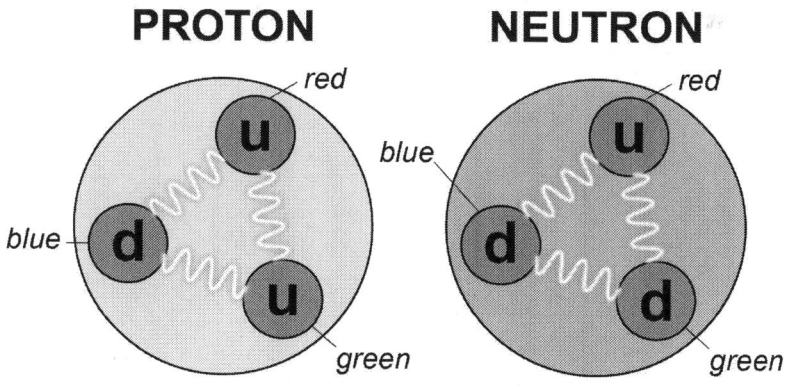

Fig. 21: Quark composition of protons and neutrons.

Before diving headfirst into these head-scratchers, let's take a step back. It's time to really understand what we mean when we say "state of matter." If we lean on our school-taught knowledge, we can identify a pretty straightforward trend: temperature changes can make matter shape-shift between different states.

These shape-shifting transitions, or *phase transitions* as we like to call them, have their own set-in-stone rules. Take water, for example. When you dial up the temperature above 273 Kelvin (equivalent to 0 degrees Celsius), ice decides it's had enough of being solid and slips into its liquid form. Crank the heat up to 373 Kelvin, and the water transforms into gas. And when things get really hot—several thousand Kelvin hot—we cross into the dazzling domain of plasma.

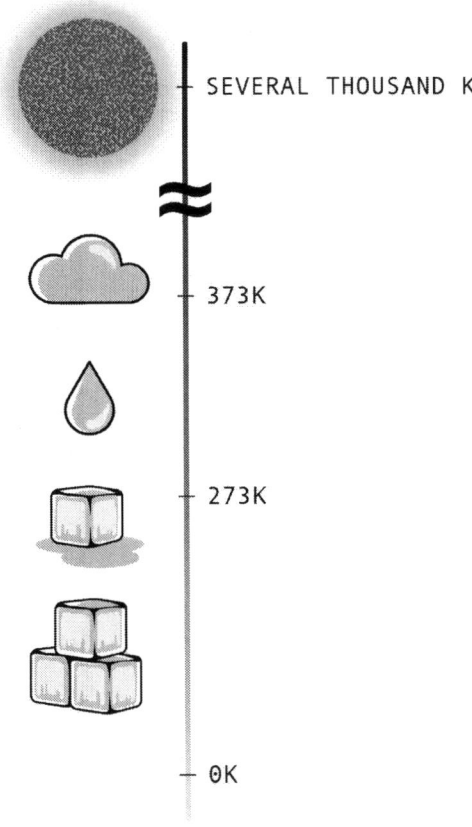

TEMPERATURE

Fig. 22: Temperature thresholds in Kelvin for transitioning between different states of matter.

But let's not oversimplify things. Reality is much more complex than "heat up, change state." These transition temperatures aren't just dependent on the type of matter we're dealing with but also on pressure. Picture yourself making a cup of tea on a mountaintop, and you'll find that water boils at a lower temperature due to the lower air pressure. So, our understanding of matter's magic tricks needs a bit of a tweak—it's not just about temperature, but pressure too.

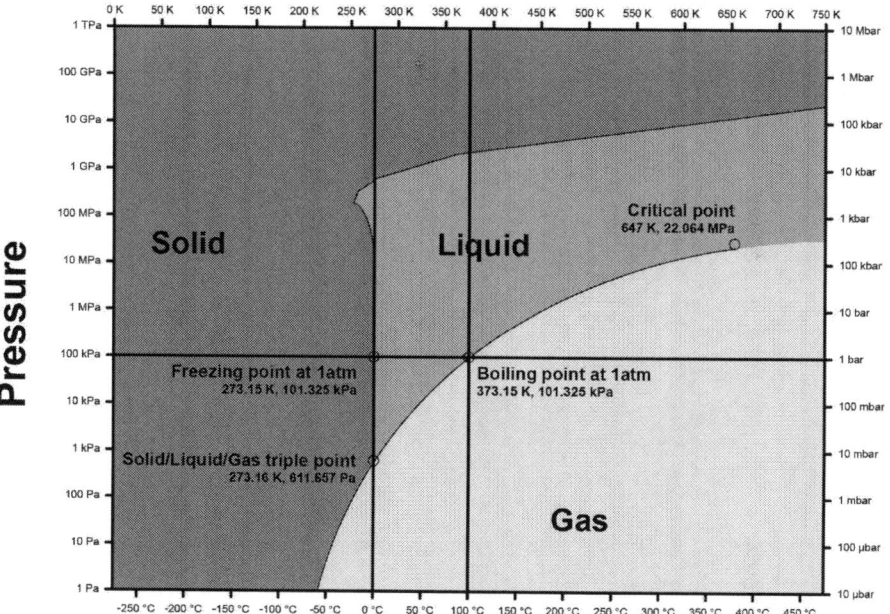

Fig. 23: Phase diagram.

Let's examine this graph, known as a *phase diagram*. It serves as a guide to the complex world of matter, offering insights far beyond the basic states of solid, liquid, gas, and plasma.

Here's a fascinating twist: when both temperature and pressure are increased beyond a certain threshold, known as the *critical point*, the distinction between gas and liquid begins to blur. This is akin to the indistinct line where the sky blends with the sea on the horizon. In this ambiguous region, a remarkable state of matter emerges, known as a *supercritical fluid*. This state defies conventional categories—it's neither a gas nor a liquid but a unique hybrid exhibiting characteristics of both.

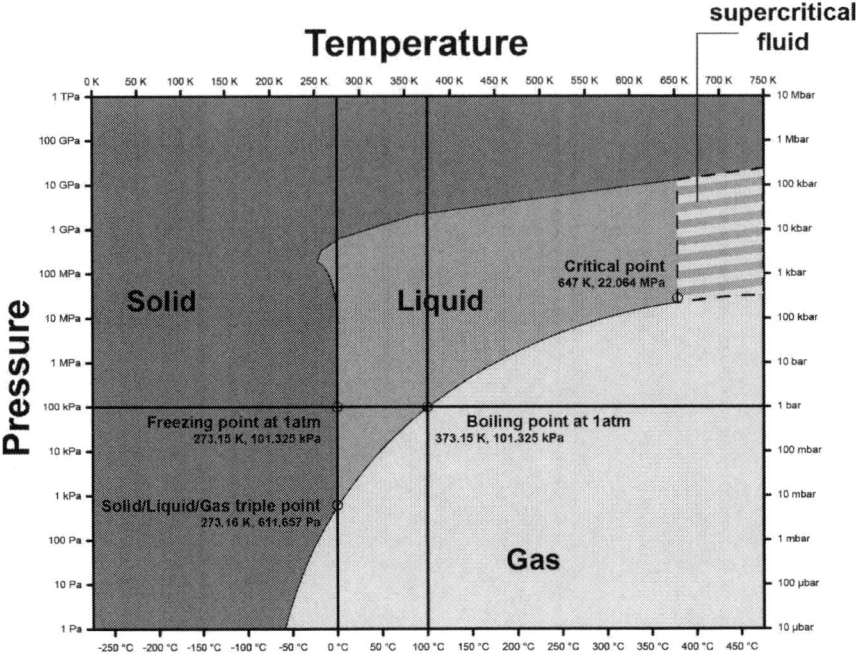

Fig. 24: Phase diagram with supercritical fluid area.

Keep in mind, when we refer to temperatures and pressures in this phase diagram, we're discussing the collective actions of a vast number of particles. It's comparable to observing a busy city from above. While a single water molecule doesn't possess a "temperature" in itself—its movement is more accurately described as speed—when you consider a large group of these molecules together, their average activity is what we define as temperature. This is how we measure the overall energy and movement of particles in a substance.

A state of matter is really about how these averages—temperature, pressure, and others—coexist and play off each other. This forms something we call the *equation of state*. Thermodynamics helps us understand these relationships between the statistical properties in different states of matter.

But don't be fooled into thinking it's all just about thermodynamics. Once we dive into the subatomic world, things get even more fascinating. We can twiddle and tweak matter into different states using more than just temperature and pressure.

Consider this hypothetical scenario: What do you think would occur if we significantly increase the temperature of plasma to an extraordinarily high level?

Inside plasma, you've got a party of particles, with electrons wandering freely while their parents nuclei, made up of protons and neutrons, stay intact. But what if we dared to turn up the heat even more? What boundaries could we push then?

Let's delve into the details: dismantling the nuclei of atoms is a monumental task. The force binding these protons and neutrons is immensely powerful, necessitating an enormous amount of energy to disrupt it. And by enormous, we're referring to a staggering 7 trillion Kelvin! This specific threshold is so significant that it has been named the *Hagedorn temperature*, a critical point in the realm of particle physics.

Just like breaking the sound barrier, reaching this temperature is a significant threshold in the realm of matter.

When we surpass the Hagedorn threshold, things take a dramatic turn. The nucleons, which are the protons and neutrons, break apart. This leads to the emergence of an exotic new state of matter known as *quark-gluon plasma*. However, it's important to temper the excitement; we can only produce minuscule quantities of this plasma in laboratory settings using particle accelerators, which act like tiny versions of cosmic factories.

However, during the nascent stages of our universe, the entire cosmos was believed to exist as a quark-gluon plasma. Furthermore, it is speculated that the cores of massive neutron stars might also harbor this extraordinary state of matter.

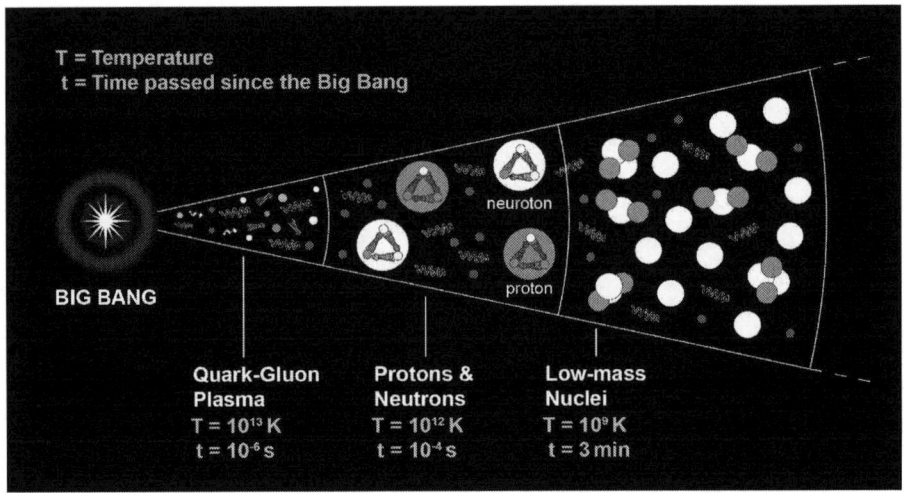

Fig. 25: Depicts the various stages of the universe's evolution, starting from the Big Bang and the creation of the quark-gluon plasma, progressing onwards.

It's intriguing to note that quark-gluon plasma exhibits certain liquid-like behaviors. However, there's even more to this fascinating state of matter. It has the ability to freeze!

The concept of "freezing" in the context of *quark-gluon plasma (QGP)* is quite distinct from the common understanding related to temperature changes in everyday substances. In general terms, freezing implies a transition from liquid to solid as temperature decreases. However, in the specialized field of high-energy physics, especially when discussing QGP, the term takes on a different meaning.

The term "freezing" in relation to QGP refers to its phase change from this energetic plasma state to a state where quarks and gluons reassociate to form regular particles like protons and neutrons. This transition, often called "freezing," is essentially the reformation of quarks and gluons into structured particles as the QGP cools from its extraordinarily high temperatures.

When it freezes, the quark-gluon plasma morphs into something we call a *hadron*. These are subatomic particles made of two or more quarks held together by a super-strong nuclear force.

Neutrons and protons are examples of these, and as you probably know, they are found inside the nucleus of every atom.

Additionally, there exist exotic combinations of quarks that are also considered as hadrons.

To better grasp the spectrum of matter states, let's imagine a hadron as a sort of crystal emerging from the quark-gluon plasma. It's almost like the solid version of this state of matter. So, in a whimsical way, we're all constructs of "quark snowflakes!"

This insight leads us to affirm that quark matter has its own distinct states, its own landscape of existence. We can depict these states in an alternative phase diagram, trading pressure for what we call *baryonic potential*. This fresh angle gives us a richer understanding of the captivating behaviors and transformations within the world of quark matter.

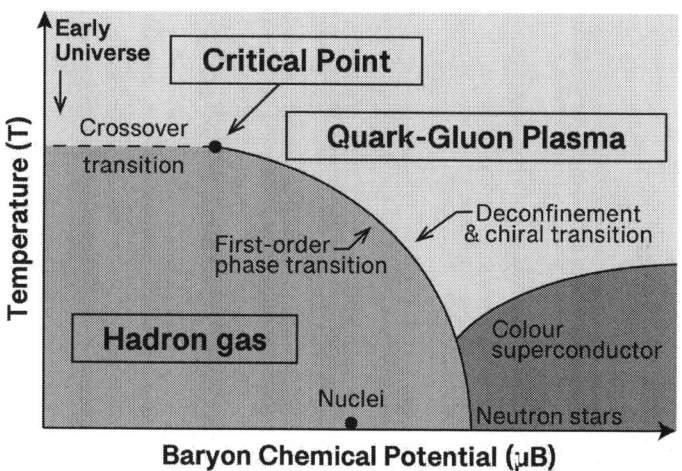

Fig. 26: *Quantum Chromodynamics (QCD) Phase Diagram.*

If the term *Baryonic Chemical Potential* is making you scratch your head, don't sweat it! It's just a fancy way of saying how much energy quarks can soak up or spit out.

Let's take another look at our phase diagram. Here, we see that our quark-gluon plasma acts a lot like gas in the world of atomic matter, even though it behaves more like a liquid. Meanwhile, Hadron gas lines up more with what we'd call a solid state.

If we stroll to the right side of this diagram, we'll see a sneak peek of what's going on inside a neutron star. Here, those "quark crystals" start to come together, forming a fluid-like stuff we call *neutronium*. Then, the neutrons within this neutronium dissolve, resulting in a special, liquid-like form of quark matter.

Fun fact—neutronium is a term coined by Andreas von Antropoff way back in 1926. He suggested there could be a type of matter made entirely of neutrons, with no protons or electrons to be found. In his view, neutronium was an element with an atomic number of zero, leading the charge in his revamped periodic table.

Let's go back one year. In 1925, a bright young physicist from India named Satyendra Bose sent a paper to none other than Albert Einstein. He'd been unable to get it published. Bose had gone on a bit of a daring adventure—he'd taken the mathematics used for light particles and tried applying them to whole atoms. Einstein realized how important this could be. He dug in deeper, did some calculations, and predicted an entirely new state of matter—one that would dance to the rhythm of quantum rules, not behaving like your everyday solids, liquids, or gases. They named this wondrous state the *Bose-Einstein condensate*, in honor of its imaginative architects.

For the next seven decades, creating this condensate was a thing of dreams. It was like the Yeti of the physics world—rumored but never seen.

The first experimental evidence of Bose-Einstein condensates was achieved in 1995 with ultracold rubidium and sodium atoms. Scientists discovered that when you go to ultra-cold temperatures, the quantum characteristics of atoms start to take center stage. At these temperatures, particles like atoms start behaving according to the principles of quantum mechanics rather than classical physics. Each atom begins to behave more like a wave and less like a speck of matter—think wave packets zipping around rather than mere dots. As for why this happens, well, that's a mystery we're still trying to unravel.

As the temperature continues to decrease, these wave packets become increasingly elongated. Then, a phenomenon occurs: when they become cold enough, these packets start to overlap. It is at this point that the system undergoes a profound transformation. The particles within the system experience what can be described as an "identity crisis." They lose track of their individual locations and merge together in a vast quantum state. All the particles, previously separate entities, become part of a singular, collective quantum system. This concept challenges our conventional understanding of physics, making it quite difficult to visualize and comprehend.

This transition marks a significant milestone in the realm of quantum mechanics. It opens up new avenues in quantum physics, setting the stage for further exploration. Scientists can now investigate the mysterious behavior of particles at near-zero temperatures more closely. Additionally, this allows for the study of quantum phenomena on a macroscopic scale.

Building on the understanding of phenomena like Bose-Einstein Condensation, which allows for the study of quantum behaviors on a macroscopic scale, we now turn our attention to another intriguing concept in quantum mechanics: *time crystals*.

These represent a peculiar quantum state of matter and are the last topic we'll discuss in this section. A time crystal refers to a configuration of entangled particles that oscillate between different states, even in the absence of traditional forms of energy. *Entanglement* is a unique quantum phenomenon where particles become interconnected in such a way that the state of one directly influences the state of another, regardless of distance.

In conventional thermodynamics, the lowest energy level corresponds to absolute zero temperature, where particles cease all motion. However, what sets time crystals apart is that their lowest energy state involves genuine motion. This makes them thermodynamically distinct from other states of matter.

These intriguing insights lead us to a few remarkable conclusions. Firstly, the concept of a state of matter extends beyond just atoms and encompasses subatomic particles as well. Subatomic particles can possess their own distinct states of matter. Furthermore, it's possible for two entirely different states of matter to coexist simultaneously at different scales. For instance, within liquid water, numerous tiny pockets of frozen, solid nuclear material can exist.

If we consider the possibility of atoms and subatomic particles having their own state of matter at different scales, it prompts the question: Does this extend to larger scales as well?

To explore this idea, let's contemplate a few grains of sand. Individually, each grain of sand is solid. However, when we allow air to flow through the sand, something remarkable happens. The interaction between the grains of sand changes, and collectively, the sand begins to exhibit behavior reminiscent of a liquid. In this state, if we place lightweight objects at the bottom of a sandbox, they will effortlessly rise to the surface. This unique characteristic sets it apart from a typical solid.

What's intriguing is that even though each grain of sand remains solid and the air used in the experiment retains its gaseous nature, the sand as a whole displays a distinct behavior. This showcases the fascinating complexities that can arise when different components interact.

Now, let's shift our perspective to a grander scale—galaxies. Astrophysicists often employ a modeling approach where galaxies are envisioned as a fluid composed of stars. The interactions within this "galactic fluid" are governed not by electromagnetic forces but by the force of gravity. Galaxies can be thought of as vast assemblies of stars, which themselves are made up of plasmas consisting of hydrogen and, intriguingly, potentially containing frozen clusters of quark matter.

Considering the behaviors observed in systems like sand and galaxies, one might wonder if these phenomena can be regarded as actual states of matter. Technically speaking, they do not fit the traditional definitions of states of matter. However, this distinction is largely a matter of convention.

The concept of a state of matter proves invaluable in helping us comprehend complex interactions, even among macroscopic "particles" like galaxies or grains of sand. Scientists have recognized the intriguing analogies between the quantum realm and the macroworld when discussing states of matter. Max Tegmark, a visionary physicist from MIT, took this concept a step further in 2014 by postulating the existence of a state of matter akin to solids, liquids, or gases. In this hypothetical state, atoms are organized in a manner that enables the processing of information and gives rise to subjectivity and, ultimately, consciousness.

While these ideas remain speculative and are subject to ongoing scientific exploration, they highlight the power of the concept of states of matter in aiding our understanding of intricate interactions at various scales—whether in the quantum realm or on a grand cosmic stage.

CHAPTER 9

Harnessing the Power of Antimatter

Antimatter, quite literally, is the complete opposite of matter. You see, for every sub-atomic particle we know, like electrons, protons, and neutrons, a corresponding antiparticle exists: the antielectron, antiproton, and antineutron, respectively. Now, here's where it gets interesting. While antiparticles share the same mass as their corresponding particles, they possess opposite charges and differ in other quantum properties.

These quantum properties are associated with various characteristics of particles. For example, we have the *lepton number*, which applies to particles like electrons and the other members of the lepton family. Additionally, we have the *baryon number*, which is one-third for each of the six quarks composing the baryon family.

Now, if you are not familiar with the baryon and lepton numbers, don't worry.

The baryon number serves as a distinctive identifier, akin to a signature unique to *baryons*. Baryons are fascinating particles that constitute the material we recognize as matter, with protons and neutrons being quintessential examples. Each baryon, whether it's a proton or a neutron, is assigned a specific marker known as the baryon number. This is essentially their official stamp confirming their baryonic identity. So, whenever you encounter a baryon, it's as if it carries its own identification card. For typical baryons, this number is designated as +1, making it a straightforward way to identify them.

However, there's an interesting addition to this concept. Non-baryonic particles, which are different from baryons, do not receive this special identification. As a result, their baryon number is zero. It's as if they were absent from the "Baryon Club" gathering. Nonetheless, these non-baryonic particles have their own unique characteristics and roles in the universe.

Here's where it gets really interesting. In the wild world of particle interactions, baryon numbers have a rule—they like to be conserved. In any particle interaction, the total baryon number of the initial particles has to be the same as the total baryon number of the final particles. It's like they're keeping the baryon number balance in check.

So, if you start with some protons and neutrons that have a baryon number of 1 each, and they decide to interact and do a little particle dance, the sum of their baryon numbers should remain the same in the end.

What about leptons, then? Leptons are particles that include the familiar electron, as well as its cousins, the *muon* and the *tau*. Each lepton carries its own unique lepton number, which is like its personal lepton ID card. It's a special mark that distinguishes them from other particles. For every lepton, like the *electron*, *muon*, or *tau*, there's a specific lepton number assigned to it. And what's that number? It's just +1 for each lepton.

Antileptons, which are the antimatter versions of leptons, also possess lepton numbers, but with a notable difference. Instead of having a lepton number of +1 like their matter counterparts, antileptons are assigned a lepton number of -1. This is akin to them being the intriguing, mirror-image counterparts of leptons, each with its own distinct identity within the particle family.

When particles interact, their lepton numbers like to play a special role. You see, lepton numbers also have their own conservation rule in particle interactions.

In any interaction, whether it's a particle collision or a transformation, the total lepton number of the initial particles must be equal to the total lepton number of the final particles.

Now, let's go back to the main topic: antimatter!

When it comes to interactions, antiparticles are expected to behave just like their particle counterparts. The laws of physics exhibit (almost) symmetrical behavior between matter and antimatter. In fact, if our universe were entirely composed of antimatter, it would be virtually indistinguishable from the one we inhabit!

Let's dive a little deeper into the fascinating world of particles and their antiparticles. One such example is the antielectron, also known as a *positron*. As the name suggests, it carries a positive charge, and its lepton number is -1. Remarkably, the positron shares the same mass as an electron.

The proton, with a baryon number of 1, forms an interesting counterpart to the antiproton. The antiproton has an equal mass to the proton but possesses a negative charge and a baryon number of -1. This negative baryon number is contributed by the presence of three antiquarks within the antiproton.

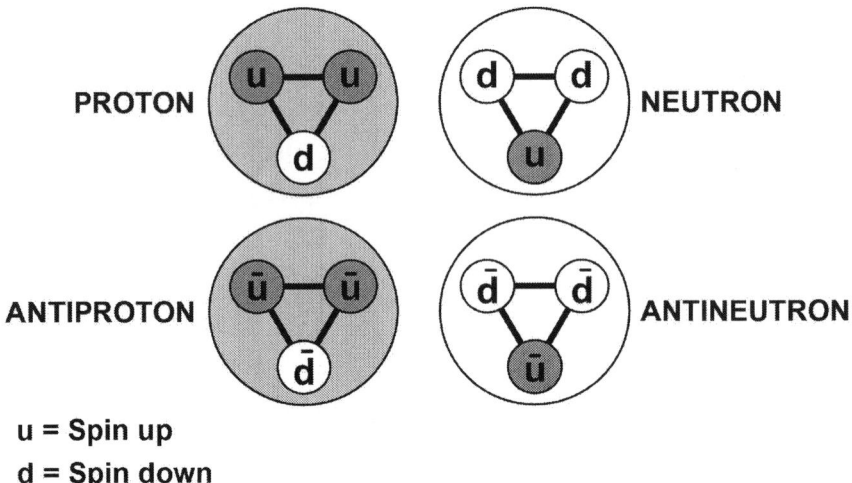

u = Spin up
d = Spin down

Fig. 27: Representation of the quark-level structures of proton-antiproton and neutron-antineutron pairs.

Now, let's take a closer look at neutral sub-atomic particles like the neutron. In the antimatter realm, we encounter its counterpart, the antineutron. The antineutron shares the same mass and zero charge as its neutral matter counterpart, but it carries a baryon number of -1. Similar to the antiproton, this negative baryon number arises from the three antiquarks within the antineutron.

Now, what happens when matter and antimatter collide? The result is truly explosive! When a positron and an electron come into contact, they undergo a process called *annihilation*, resulting in the production of two photons with X-ray energy. Thankfully, antimatter is exceptionally rare in our universe today. Otherwise, our encounters between matter and antimatter would generate a constant barrage of X-rays and gamma rays.

ANNIHILATION

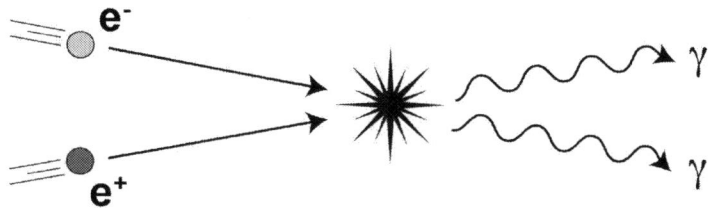

Fig. 28: Annihilation of an electron and a positron, resulting in the conse-quent release of gamma rays.

Interestingly, if one were to encounter their antimatter counterpart, both would annihilate, releasing an enormous amount of energy. The energy released from such an encounter would be equivalent to roughly 2500 megatons, which is nearly one-third of the total energy stored in the world's arsenal of nuclear weapons!

Antimatter holds tremendous potential for applications like power-ing spaceship engines or even supplying energy to entire nations. However, there's a catch. To harness this power, we must overcome the significant challenge of generating substantial amounts of mac-roscopic antimatter and storing it securely in isolation.

Currently, modern particle accelerators have the capability to pro-duce antiprotons in small quantities for use in high-energy physics experiments. However, we're dealing with minuscule amounts here. In order to unlock the true potential of antimatter as a practical en-ergy source, we need to develop techniques to generate and accumu-late significantly larger quantities.

Now that we know what antimatter is, let's see how physicists dis-covered it. Our journey takes us back to the year 1928. This was when Paul Dirac, a visionary figure and one of the founding fathers of quantum mechanics, grappled with a significant challenge. He was working on an equation that sought to incorporate the effects of Einstein's theory of special relativity into the description of electron behavior in the microscopic realm.

To Dirac's astonishment, his equation yielded solutions that suggested the existence of electrons with negative energy, seemingly moving backward in time! An inexperienced physics student might have been overcome with embarrassment and swiftly recalculated their math. However, Dirac, with unwavering confidence, believed in the validity of his calculations. Rather than discarding the perplexing result, he bravely reinterpreted it as the existence of an *antielectron* possessing positive energy, progressing forward in time.

Fast forward four years, and enter Carl Anderson, an experimental physicist who would play a crucial role in validating Dirac's audacious hypothesis. In 1932, Anderson made a groundbreaking observation: he actually detected the presence of the predicted antielectron, which came to be known as the *positron*. This groundbreaking discovery earned Anderson the prestigious Nobel Prize in 1936, while Dirac had already been recognized with a Nobel Prize in 1933 for his remarkable contributions to the field of atomic physics.

Do you want to know about the origin of the positrons discovered by Anderson? Anderson's positrons were generated in atmospheric showers of subatomic particles. These showers are the results of cosmic rays, mostly composed of protons, colliding with atoms in our planet's atmosphere.

At the time, Anderson employed a device called a *cloud chamber*, a common instrument in nuclear physics laboratories. When a particle traverses through a cloud chamber, it leaves a trail of visible bubbles in its wake. By applying a known magnetic field, the charged particle would curve accordingly. Fascinatingly, Anderson observed that the positron in his cloud chamber curved in precisely the same manner as electrons but in the opposite direction.

It's worth noting that several years prior to Anderson's breakthrough, other physicists had actually observed these tracks curving in the opposite direction, similar to the positron's trajectory. However, none of these scientists pursued the matter further. It was Anderson's unwavering determination that set him apart. He meticulously tested and retested his results, ultimately confirming the existence of the positron. This mirrors Dirac's approach when confronted with negative energy electrons. Rather than discarding them as mathematical oddities, Dirac reinterpreted the solutions, ultimately leading to the discovery of an entirely new aspect of the universe.

The scarcity of antimatter in our present-day universe can be attributed to its formation through exceptionally rare nuclear reactions, yielding only minute quantities. The antimatter we manage to obtain primarily exists at the sub-atomic level, encompassing particles such as positrons, antiprotons, antineutrons, and antimesons. Mesons, in particular, serve as the lighter counterparts to protons and neutrons, consisting of a quark paired with an antiquark.

Physicists have indeed created and studied antihydrogen. However, this remarkable feat was accomplished under challenging laboratory conditions, necessitating the meticulous isolation of antihydrogen atoms to prevent their interaction with container walls. Such interactions would result in annihilation, releasing gamma rays.

Now, you may be wondering whether antimatter, in our modern world, serves any practical purposes beyond expanding our understanding of the universe. The answer is a resounding yes! Antimatter plays a crucial role in a medical procedure known as *Positron Emission Tomography*, or *PET scan*, which aids in the detection of tumors within the human body.

During a PET scan, a patient is administered radioactive sugars that specifically accumulate at tumor sites. Over time, these radioactive sugars undergo a process of gradual decay, emitting positrons along the way. In this crucial stage, a specialized and exceptionally advanced instrument precisely maps out the origin of these emitted positrons, creating an intricately detailed visualization of the interior of the human body.

Through this innovative technique, medical professionals gain invaluable insights into the presence and precise location of tumors, enabling more accurate diagnoses and tailored treatment plans. PET scans empower healthcare providers to visualize the invisible, allowing for earlier detection and improved management of various medical conditions.

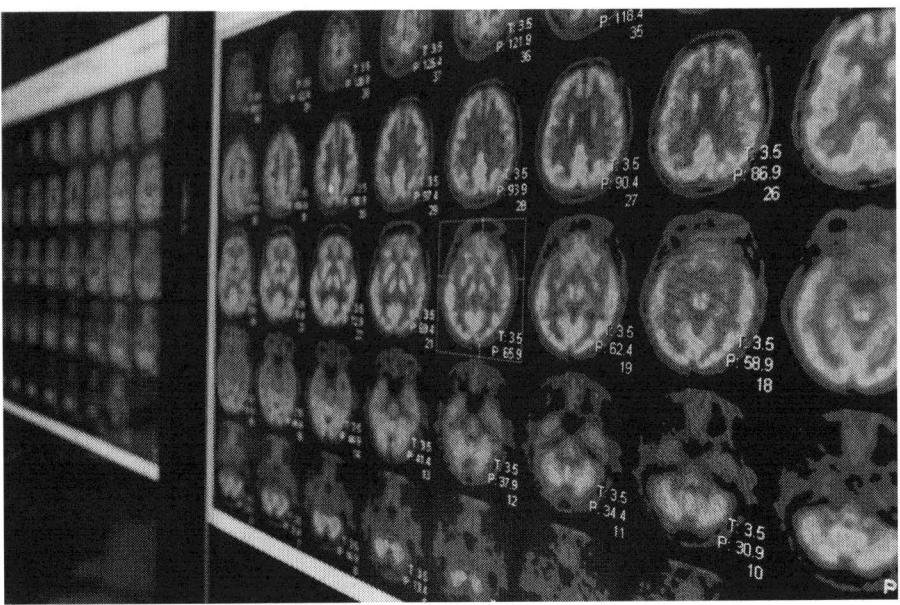

Fig. 29: PET ct scan of human brain.

Having explored the intricacies of antimatter—its nature, interactions with matter, production mechanisms both on Earth and in nature, and its transient existence due to rapid interactions with ordinary matter—we arrive at a compelling question: Why does the vast majority of the universe, encompassing planets, stars, and galaxies, consist predominantly of matter rather than antimatter? This question holds such profound significance in the realm of physics that it has acquired its own distinctive name: the *matter-antimatter asymmetry problem*.

Exploring the matter-antimatter asymmetry problem takes us into the realms of cosmology and high-energy physics, also known as particle physics. It's important to note that while physicists are actively working on finding a definitive solution, the puzzle is not yet fully resolved. Nevertheless, we can provide a simplified overview of the current understanding surrounding this fascinating issue.

To comprehend the matter-antimatter asymmetry, we delve into a theoretical framework formulated over fifty years ago by Andrei

Sakharov. Our journey commences at the very genesis of the universe—the Big Bang. This monumental event marked the birth of our cosmos, and during this primordial phase, equal amounts of matter and antimatter were generated. Each electron had a corresponding positron, and every quark had an antiquark counterpart. However, an intriguing twist unfolds.

Protons and neutrons, which constitute the building blocks of matter as we know it, emerged a few minutes after the Big Bang. By that time, most of the antimatter had already undergone annihilation, leaving behind a predominance of matter. Now, let's rewind and return to the epoch of the Big Bang itself.

In the initial minuscule fraction of a second following the Big Bang, the universe existed as an intensely hot amalgamation of particles, antiparticles, and pervasive radiation. During this epoch, particles and antiparticles constantly collided, annihilating each other and generating highly energetic photons. In turn, these energetic photons could interact and spawn new pairs of particles and antiparticles, governed by the fundamental laws of particle physics.

This fiery equilibrium would have persisted indefinitely if not for a remarkable cosmic phenomenon: the expansion of the universe. As the universe expanded, it underwent cooling. The decreasing energy of radiation eventually reached a point where photon interactions could no longer generate particle-antiparticle pairs with sufficient energy. Consequently, the production of these pairs ceased, and they became rare.

If no additional processes had intervened, particles and antiparticles would have eventually annihilated one another, leaving behind a universe solely composed of photons, devoid of matter and antimatter. In such a scenario, we would not be here engaging in this discussion.

However, the expansion of the universe played a pivotal role in disrupting this self-annihilation cycle. By halting the production of particle-antiparticle pairs, it set the stage for other processes to come into play, marking the second prong of our exploration. It's important to recall that antiparticles possess opposite baryon numbers compared to particles.

These additional processes, rooted in high-energy physics, must adhere to the conservation principles of energy, momentum, charge, lepton number, and baryon number. They play a crucial role in establishing the matter-antimatter asymmetry.

In the realm of high-energy physics, the concept of conservation is fundamental. A quantity is considered conserved if its value remains the same before and after a reaction takes place. The laws governing high-energy physics demand strict conservation of various quantities. However, the intriguing absence of antimatter in the universe suggests that, under specific conditions, the conservation of baryon number must be violated to some degree in order to explain the observed matter-antimatter asymmetry.

To quantify this violation, scientists estimate that it occurs at a minute level of approximately one part in one billion. How do we arrive at this estimation? Astrophysicists have discerned that for every particle in the universe, there are roughly a billion photons present. This implies that out of a billion particles, only a billion antiparticles are observed. Consequently, when the billion particles annihilate with the billion antiparticles, approximately a billion photons are produced. Consequently, we are left with a surplus of one particle and a billion photons, aligning with the particle-to-photon ratio observed by astrophysicists.

However, it is crucial to note that thus far, no experimental evidence of baryon number conservation violation has been detected in particle physics laboratories. Moreover, within the accepted framework of particle physics, known as the *Standard Model*, there is no theoretical process in which the baryon number is not conserved. This realization emphasizes that achieving a comprehensive understanding of the matter-antimatter asymmetry necessitates the existence of new physics that extends beyond our current knowledge.

The third crucial aspect in understanding the matter-antimatter asymmetry is known as *CP violation*. In the realm of physics, the laws, including conservation laws, are expected to remain unchanged if we invert the signs of charges (C) for all particles involved in a reaction and view the reaction through a mirror (P: Parity). However, CP violation occurs when the outcomes of a reaction change upon the reversal of charge signs and the mirror reflection of the reaction.

Since the 1980s, glimpses of CP violation have been observed in particle physics laboratories, particularly in reactions involving K-mesons (also known as *kaons*) and other subatomic particles. However, the degree of observed violation is not substantial enough to fully account for the magnitude of the matter-antimatter asymmetry. Further experiments are still required to unravel the complete picture and unveil the underlying mechanisms driving this asymmetry.

In summary, it is indeed true that a small quantity of antimatter has the potential to generate a substantial amount of energy, capable of powering a city for several days. However, our current technological capabilities are not advanced enough to efficiently produce and harness antimatter for energy generation through annihilation. Streamlining the production process and effectively utilizing the energy released remain significant challenges.

Antimatter has captivated physicists for nearly a century, providing profound insights into the workings of the universe. While we have made significant strides in understanding antimatter, its rarity continues to present a captivating puzzle. Investigating the reasons behind its scarcity holds the potential to unlock new frontiers of physics and expand our understanding of the cosmos. This ongoing saga unfolds within major particle physics laboratories worldwide, with institutions like CERN in Geneva, Switzerland, at the forefront of this pursuit. Immense dedication and collective efforts are directed toward unraveling the intricacies of the matter-antimatter asymmetry.

CHAPTER 10

Is Teleportation Possible?

The concept of teleportation has captivated our imagination through various works of science fiction. But what about its feasibility in reality? Teleportation refers to the hypothetical transfer of matter or energy from one location to another without physically traversing the space in between. While modern science has yet to achieve the actual teleportation of matter, it remains a topic of speculation.

The challenge lies in the fact that any transfer of matter without physically moving through space violates Newton's laws of motion. This raises doubts about the practical realization of teleportation in the traditional sense. However, in the realm of quantum mechanics, a different form of teleportation becomes possible.

Quantum teleportation involves the manipulation of entangled particles that are separated by vast distances. These particles exhibit a special connection, known as *entanglement*, that enables information to be instantaneously transmitted between them. By utilizing a third particle as a messenger, information about the state of one entangled particle can be transferred to the other, effectively teleporting the quantum state.

Let's begin by understanding what it means for two particles to be entangled. In the world of quantum mechanics, every particle has a set of properties, such as its position, energy, momentum, and spin, which are described by a mathematical function called the *wave function* or *quantum state function*. This wave function holds all the information about the particle.

When we measure one of these properties of a particle, we gain some insight into its overall quantum state as encoded in the wave function. However, according to the quantum uncertainty principle, this act of measurement disturbs the other properties of the particle, making it impossible to simultaneously determine them precisely. Essentially, the act of measurement alters the pristine state of the particle and destroys some of its original information. This is why we cannot simply copy particles and teleport their exact states to another location through a process called *quantum teleportation*. Nevertheless, there is an intriguing possibility: we can recreate an unmeasured quantum state in a different particle, but it comes at the cost of sacrificing the original particle.

What I'm going to explain is one of the weirdest characteristics of quantum mechanics, which was revealed back in the 1930s.

Einstein, Podolsky, and Rosen discovered in 1935 that two quantum subsystems, even when separated in space, can be 'entangled' in a way that defies our everyday understanding of cause and effect. When these particles interact with each other, they can become entangled. In this peculiar state, both particles remain connected as part of the same quantum system, so any action performed on one of them instantly affects the other in a predictable manner.

To clarify further, when we measure one of the entangled subsystems, it enters a specific quantum state, and astonishingly, the entangled subsystem that is separated by large distances in spacetime (yes, even backward in time!) also instantaneously assumes a corresponding quantum state. This non-local and seemingly instantaneous correlation between entangled particles is truly puzzling and represents a fascinating aspect of quantum mechanics.

In quantum mechanics, particles possess a property called *spin*, which can be either 'up' or 'down.' However, in the quantum realm, particles can exist in a superposition of both states simultaneously.

Let's imagine we have two entangled particles placed in two different cities. This entanglement could result in a peculiar scenario where both particles are simultaneously in a superposition of spinning up and spinning down.

Now, here comes the interesting part. If we measure the spin of one of these particles, let's say the one in New York, we will obtain a definite measurement, either up or down. The surprising aspect of quantum entanglement is that as soon as we make this measurement and find, for instance, that the particle in New York is spinning up, we instantly know that the particle in Los Angeles is spinning down.

What's truly intriguing is that these particles could be separated by vast distances, yet they seem to communicate with each other instantaneously, defying our classical intuition. Albert Einstein referred to this phenomenon as "spooky action at a distance."

It's important to note that this instantaneous correlation between the measured spin of one particle and the known spin of the other particle doesn't involve any physical interaction between them after they become entangled. Instead, it's a manifestation of the intricate interconnectedness of quantum mechanics.

While we currently lack the ability to physically transport matter from one place to another, the intriguing aspect of quantum entanglement is that it allows for the transfer of information in a way that seems almost like teleportation.

In 2017, a group of dedicated Chinese scientists achieved something remarkable: they successfully demonstrated quantum entanglement through the expanse of space. If this sounds like something straight out of a science fiction novel, don't worry: let's break it down.

This highly sophisticated experiment, conducted on a satellite named Micius, showed us that quantum states can indeed be shared over vast distances, extending as far as 1,200 kilometers. This accomplishment shattered the previous record, redefining our understanding of the quantum world.

Now, you might be thinking, "Why is this considered a significant achievement? Shouldn't there be no limit on the distance for quantum teleportation?" Well, here's the catch: entanglement, the key ingredient in this process, is incredibly delicate. When photons interact with matter in the atmosphere or within optical fibers, the entanglement can be easily disrupted or lost. This poses a major challenge and has previously reduced the distance over which scientists can reliably measure entanglement or carry out teleportation to only around 100 kilometers.

So, what's cooking inside the Micius satellite? Imagine an ultraviolet laser beam that is cleverly split and directed into a specialized crystal. This crystal becomes the birthplace of pairs of photons, entangled in a mysterious dance, their polarizations opposite of each other.

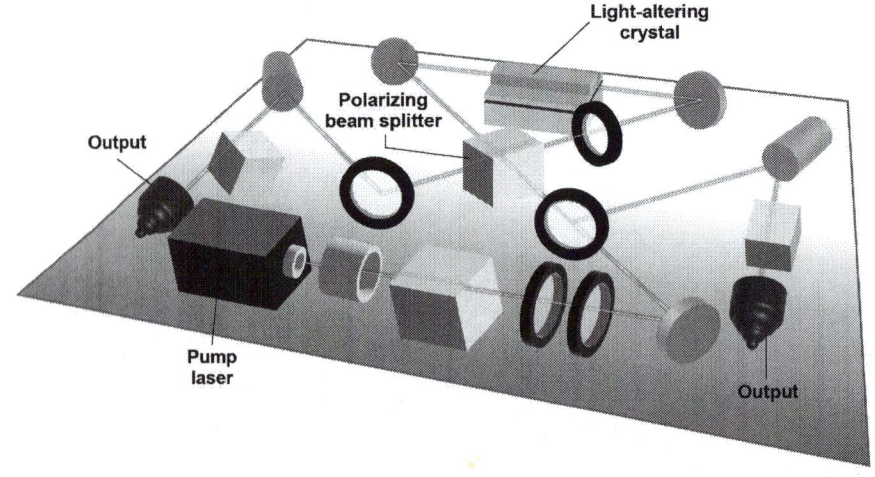

Fig. 30: Explanation of the ultraviolet laser beam mechanism within the Micius satellite.

These entangled photons then rain down to Earth at a whopping rate of 5.9 million pairs per second. They're received by two ground stations located in China, positioned over 1,200 kilometers apart.

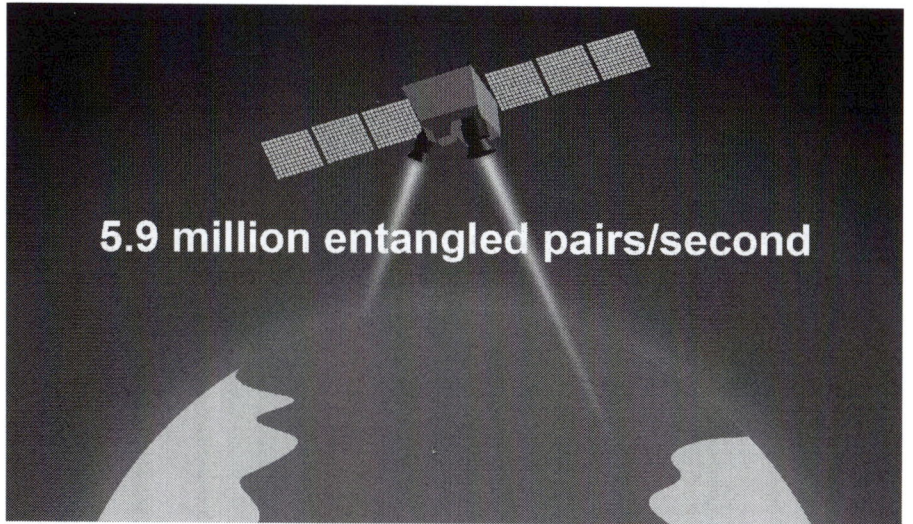

Fig. 31: Chinese satellite emitting entangled photon pairs, aimed to be received by two ground bases on Earth.

Now, the scientists at these ground stations measure the states of the incoming photons and notice something phenomenal. The states of these photons are correlated much more than you would expect by just chance—the smoking gun evidence that these photons were maintaining their entangled state even at such great distances. Yet, while this space-based demonstration of quantum entanglement was far more efficient than the earthbound ones, it wasn't flawless. Only 1 in 6 million photons made it to the ground stations.

Take a moment to ponder the possibilities that arise from the recent advancements in quantum teleportation. Imagine a future where communication transcends physical barriers and encryption becomes a thing of the past. While we're not there yet, the potential is fascinating.

In the realm of quantum physics, observing a particle alters its state. However, with the concept of quantum entanglement, we only need to observe one particle to glean information about the others. This creates an avenue for one-way data transfer. But let's be clear: we still face challenges when it comes to reliably teleporting even a single photon. So, utilizing quantum mechanics for faster-than-light communication is still beyond our reach.

However, what we can do is employ quantum entanglement as a form of secure communication, much like decoder rings from the past.

Chinese scientists are setting their sights on distributing quantum keys to these ground stations. This procedure involves long chains of entangled photons. So why are these keys so special?

Well, imagine two people at different locations being able to chat privately using these quantum keys for encryption.

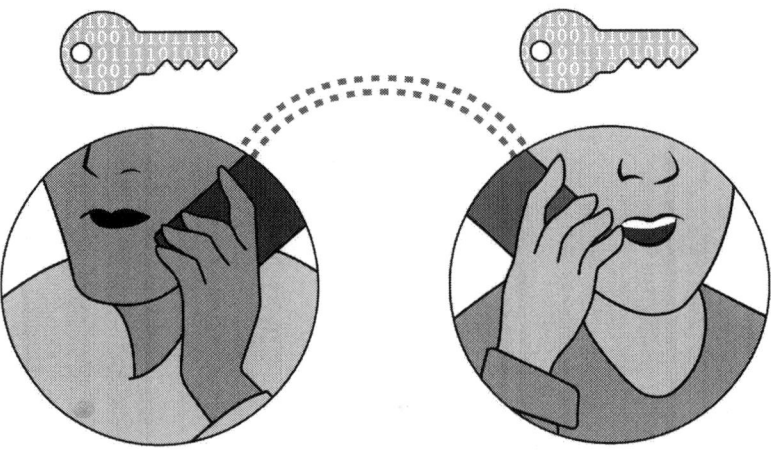

Fig. 32: Representation of an encrypted conversation.

Not only Chinese scientists but many other companies are already exploring the potential of quantum entanglement as a future security protocol for the Internet. By leveraging this phenomenon, we could potentially move data without relying on wires or wireless frequencies. And since observing entangled photons would inevitably alter their state, the level of security achieved would be truly incredible. The ultimate privacy setting, wouldn't you say?

In a significant development in 2021, a team of researchers from QuTech in the Netherlands accomplished a remarkable feat: they established the world's first multi-node quantum network by connecting three quantum processors. This groundbreaking achievement also included a proof-of-principle demonstration of essential quantum network protocols. It's important to note that this quantum network is currently in its early stages and spans a limited distance within the same building.

While we still have a long journey ahead before we can achieve true teleportation, if it is indeed possible, the technology surrounding quantum computers is set to evolve rapidly in the coming years. Advancements in quantum computing will go hand in hand with the transmission of quantum information through the phenomenon of quantum entanglement.

As we delve deeper into the realm of quantum technologies, the potential for groundbreaking advancements in communication and data processing is immense. While teleportation may remain a distant aspiration, the progress being made in the field of quantum computers holds great promise for the future of scientific exploration and technological innovation.

Conclusions

Dear Cosmic Voyager, we've reached the end of our journey. If you're here with me, we've accomplished something remarkable together. We've delved into the nucleus of reality itself, exploring the wild and wondrous world of quantum physics. We've danced with photons, bantered with antimatter, and gazed into the maws of black and white holes alike.

Through each chapter, we've sought to unravel the profound enigmas of the cosmos. We've engaged with questions that have tickled human curiosity for millennia, using the elegant language of quantum physics to decipher the universe's tantalizing secrets. We didn't need a chalkboard full of equations or a lab full of expensive equipment—just our innate curiosity and a shared sense of wonder.

We asked if it was possible to outrun light itself and pondered upon the perplexing concept of evaporating black holes. We've explored the elusive realm of antimatter and questioned the very nature of the Sun's blazing color. These quantum inquiries guided our journey through the cosmos, one page at a time.

Along the way, we have begun to see how deeply interconnected our universe truly is. How the behavior of particles billions of light-years away can be intimately tied to the ones nestled in the heart of a raindrop. We've borne witness to a world where particles can exist in two places at once, and cats can be both dead and alive—well, until someone decides to take a peek.

So, what's next for you, intrepid explorer? The universe, as they say, is your oyster. The concepts we've unraveled together, the cosmic mysteries we've probed—they're just the tip of the cosmic iceberg. There's a vast universe out there, teeming with undiscovered wonders and unexplored frontiers. Perhaps you'll set your sights on the study of dark matter or venture into the accelerating expansion of the universe.

The take-home message from our expedition, however, is not confined to quantum curiosities. I hope it is clear, now more than ever, that learning and understanding are not about complex equations or intimidating jargon. They are, instead, about nurturing a sense of wonder, about fostering an insatiable curiosity, about never being afraid to question what we think we know.

As Carl Sagan once said, "We are a way for the cosmos to know itself." And isn't that what we've been doing—knowing the cosmos, one quantum mystery at a time? Our journey may have drawn to a close, but our understanding of the universe has opened a new chapter, the pages of which are blank and ready for discovery.

In the grand cosmic scheme of things, our journey together is just one small step for a man, one giant leap for mankind. So, as we part ways, remember this: keep exploring, keep wondering, and keep gazing up at the stars. Our journey through the Quantum Horizons has shown us, if nothing else, that the cosmos is far more beautifully bizarre than we could ever have imagined.

Safe travels, dear voyager, until our paths cross again in the cosmic vastness. May you always remain curious.

Bibliography

Arvin Ash. (n.d.). YouTube. Retrieved June 27, 2023, from https://www.youtube.com/@ArvinAsh

Ashtekar, A. (2020). Black Hole Evaporation: A Perspective from Loop Quantum Gravity. *Universe*, 6(21).

Baker, D. J. (2010, March). Antimatter. *The British Journal for the Philosophy of Science*, 61(1), 93-121.

Bianchi, E., Christodoulou, M., & D'Ambrosio, F. (2018, November). White Holes as Remnants: A Surprising Scenario for the End of a Black Hole. Classical and Quantum Gravity, 35(22). 10.1088/1361-6382/aae550

Curran, M. J. (2023, March). Further considerations about light and gravity in terms of classical mechanics. *Physics Essays*. www.researchgate.net. 10.4006/0836-1398-36.1.55

Davis, E. W. (2004, August). Teleportation Physics Study. *Air Force Research Laboratory Air Force Materiel Command Edwards Air Force Base*, 40-46. https://sgp.fas.org/eprint/teleport.pdf

Dove, J., Kerns, B., & McClellan, R. E. (2021, February 24). The asymmetry of antimatter in the proton. *Nature*, 590, 561-565. https://doi.org/10.1038/s41586-021-03282-z

Dutch researchers establish the first entanglement-based quantum network. (2021, April 15). QuTech. Retrieved June 27, 2023, from https://qutech.nl/2021/04/15/dutch-researchers-establish-the-first-entanglement-based-quantum-network/

Effect of light on gravitational attraction. (2011, December). *Physics Essays*. www.researchgate.net. 10.4006/1.3653936

Einstein, A. (n.d.). Zur Elektrodynamik bewegter Korper (On the electrodynamics of moving bodies). *Annalen der Physik*, 17(891).

Einstein, A. (1912). *The theory of relativity*. George Braziller.

Exploratorium. (n.d.). YouTube. Retrieved June 27, 2023, from https://www.youtube.com/@exploratorium

First Object Teleported from Earth to Orbit. (2017, July 10). MIT Technology Review. Retrieved June 27, 2023, from https://www.technologyreview.com/2017/07/10/150547/first-object-teleported-from-earth-to-orbit/

Frieman, J. (n.d.). Fermilab. YouTube. Retrieved June 27, 2023, from https://www.youtube.com/@fermilab

Insane Curiosity. (n.d.). YouTube. Retrieved June 27, 2023, from https://www.youtube.com/@InsaneCuriosity

Juan, Y., Li, Y. H., Liao, S. K., & et al. (2020). Entanglement-based secure quantum cryptography over 1,120 kilometers. *Nature*, 582, 501-505. https://doi.org/10.1038/s41586-020-2401-y

Lorentz, H. (n.d.). Electromagnetic phenomena in a system moving with any velocity smaller than that of light. *Proceedings of the Academy of Sciences Amsterdam*, VI(809).

O'Dowd, M., Halevy, A., & Kornhaber, A. (n.d.). *PBS Space Time*. YouTube. Retrieved June 27, 2023, from https://www.youtube.com/@pbsspacetime

Poincaré, H. (1504, June 5). Sur la dynamique de l'electron (On the dynamics of the electron). *Comptes rendus de l'Académie des Sciences*.

Rielli, D. (n.d.). *Amedeo Balbi*. YouTube. Retrieved June 27, 2023, from https://www.youtube.com/@AmedeoBalbi

Sabine Hossenfelder. (n.d.). YouTube. Retrieved June 27, 2023, from https://www.youtube.com/@SabineHossenfelder

Scarmato, T. (2023, May 16). The Speed of Light in the Quantum Vacuum Aether: the "Equation of God." www.researchgate.net.

Science Time. (n.d.). YouTube. Retrieved June 27, 2023, from https://www.youtube.com/@ScienceTime24

Seeker. (n.d.). YouTube. Retrieved June 27, 2023, from https://www.youtube.com/@Seeker

Thompson, A. (2017, July 12). *Chinese Scientists Successfully Teleported a Particle to Space*. Popular Mechanics. Retrieved June 27, 2023, from https://www.popularmechanics.com/science/news/a27271/chinese-scientists-successfully-teleported-a-particle-to-space/

Unruh, W. G. (1976, August 15). Notes on black-hole evaporation. *Phys.* Rev., 14(870).

Veritasium. (n.d.). YouTube. Retrieved June 27, 2023, from https://www.youtube.com/@veritasium

Yu-Zhu, C., Yu-Jie, C., Shin-Lin, L., & Dai, W.-S. (2021). Model of black hole and white hole in Minkowski spacetime. *The European Physical Journal C volume*. https://doi.org/10.1140/epjc/s10052-021-09901-3

Printed in Great Britain
by Amazon

53016112R00062